FOREWORD BY ERIN BROCKOVICH

UNDER CORPORATE SKIES

A STRUGGLE BETWEEN PEOPLE, PLACE AND PROFIT

MARTIN BRUECKNER AND DYANN ROSS

FOREWORD
ERIN BROCKOVICH

I first became aware of the health problems members of the Yarloop community were suffering when a resident of Yarloop emailed me in 2007. People living near Alcoa's Wagerup alumina refinery (opened in 1984) had been reporting a disproportionate level of respiratory problems, skin irritations, sore throat and eyes, extreme fatigue, mental dysfunction, stomach upset, blood noses, cancers and organ failure for more than a decade. These claims were regularly reported in the media, and were the subject of an expose by the ABC's Four Corners program in 2005.

The symptoms immediately rang a bell for me — as indeed they had for the ill woman who contacted me. They reminded me of the illnesses experienced by the people of Hinkley, California in the case of groundwater contamination that started me on the path of community and environmental activism more than a decade ago. There was also the matter of a housing buyback scheme run by Alcoa, and that made me suspicious. When industry comes in and starts buying homes, you want to look closely at what's going on.

So in my first work outside of the US, I became involved in the efforts of the people of Yarloop (and nearby Hamel and Cookernup) to get their concerns addressed by the multinational corporation, Alcoa, and also by the state government of Western Australia who, despite the report of a Parliamentary Inquiry that expressed concerns about the health issues, nevertheless approved a major expansion of the refinery in 2006. Alcoa, meanwhile, repeatedly insists that the refinery is safe for both residents and workers, citing as proof the (immaterial) fact that it is one of the most studied industrial facilities in Australia.

Vince Puccio, Yarloop resident and co-chair/spokesperson for the Community Alliance for Positive Solutions action group, has said that for the residents, 'It's about accountability and for them to take full responsibility for what they've done.' But how do you get governments and companies of this scale to be accountable to a small, local group of people for a problem that they won't even admit exists? The social and environmental costs of industrial growth are too often sidelined in favour of the financial profit that it brings.

In 2007 I visited Western Australia, though I did not go to Yarloop itself — I have been made sick myself, and lost years of work, from exposure to poisonous chemicals, and I wasn't going to put myself in the way of that danger again. Nevertheless I helped to build the legal action case and in 2009 the lawyers acted on behalf of the residents and lodged a writ with a US court on the basis that Alcoa knowingly, negligently and recklessly operated its factory, poisoning surrounding communities with toxic emissions and that they concealed the toxic dangers of their refining operations.

There are many issues at stake in this conflict between the community and the corporation, but what has always concerned me most in such situations is to expose and challenge the deceits and cover ups that end up jeopardising public health and safety. I am an advocate for awareness, the truth, and a person's right to know. I believe that without the truth, we are helpless to defend ourselves, our families and our health. On that count alone this book makes an important contribution to a more informed public in the matter of the Wagerup refinery.

But while local communities should be assured of the right to the truth, they should also be assured of the right to be heard. As people who have long dwelt in a community, and who wish to be able to continue to live there with their families and neighbours without undue fear for their health, they should have a say in decisions which affect their environment. Even the Parliamentary Inquiry found that Alcoa 'failed to adequately recognise and respond to the complaints it received from workers and the local community,' and that Alcoa and the government failed to offer an unequivocal and comprehensive response to 'a range of extremely serious and complex issues at Alcoa's refinery'.

And that is another reason why the work of Martin Brueckner and Dyann Ross in this book is so important. They set out to provide above all a platform for the voices of the community — the least powerful people in this conflict — to be heard. As the authors point out, the people of Yarloop speak from a deep-rooted local knowledge. To fail to listen to them is to shut off an important source of lived experience and information that speaks directly to the many challenges of creating sustainable industry. They know the consequences. They live with them.

It is not only the health of individuals and the natural environment that is affected, there are social impacts too. In the words of one of Yarloop's residents cited here, 'This whole town has been fragmented, it's been divided, you call it whatever you want, but it doesn't even have 10 per cent of what we used to have as a community, and we had a very strong community here.'

Well, events are still unfolding at Yarloop, and I think that this book will play some part in strengthening the sense of morale and community among the people whose lives are still on the line.

Erin Brockovich
May 2010

A peaceful rural setting (Photo by H. Seiver)

CONTENTS

Foreword by Erin Brockovich	3
Preface	8
Acknowledgements	10
Introduction	12
1 A small town and its corporate neighbour	18
2 The pathology of industry-community relations	37
3 Local stories about regional (un)sustainability	61
4 The corporation's public story	126
5 Small government and big business	159
6 Sustainability doublespeak	199
7 Towards a just ethic of people, place and profit	238
Notes	271
References	294
Index	312

PREFACE

This book tells the story of a small Australian country town and its struggle with its corporate neighbour and the government. Even though it deals with a particular community, a particular company and a particular government in a discrete location, this tale could be told almost anywhere in Australia where communities live in close contact with industry, be it in Collie, the Hunter Valley, King George Sound, the Latrobe Valley, East Gippsland or Gladstone.

Australia's quest for economic growth and development has often come at hidden social and environmental costs, with heavy industries initially largely invisible due to their small scale and relative isolation from human settlements. The growth and intensification of industrial life in Australia in recent decades has, however, brought industry literally to the doorsteps of many small communities, making increasingly apparent the trade-offs associated with industrial activity and highlighting the need for renegotiating the respective importance of people, place and profits.

Under Corporate Skies provides a community account of life with big business that is symbolic of Australia's socioeconomic landscape today. It draws attention to the questions surrounding the social and environmental sustainability of towns in the shadow of heavy industry and the roles and responsibilities of body corporates and government for the protection of the health and wellbeing of affected communities.

While Alcoa World Alumina and the Western Australian state government are implicated in the plight of the people of Yarloop and surrounding areas, the intention of this book is not to lay blame. Instead, we seek to unearth mistakes made and draw attention to the lessons that can be learned with a view to preventing future conflicts such as this. We hope that the story of the Yarloop community can serve as a reminder of the rights and interests of the intended beneficiaries of economic development and as a prompt for the critical re-evaluation of the things that really matter, locally and nationally.

Martin Brueckner and Dyann Ross

ACKNOWLEDGEMENTS

We are indebted to the support of Alcoa and Edith Cowan University for funding the research that went into this book and to Curtin University of Technology for enabling Martin to continue his work on the project after leaving ECU in 2008. We are also immensely grateful for the help and assistance we recieved from Dr Meredith Green with the, at-times, overwhelming task of data analysis. Most importantly, our thanks go to the people we interviewed, who gave us insights into their lives and experiences. The courage of many people in telling their stories made this book possible.

Members of the Yarloop community, on whose behalf this book has been written, dedicate this work to the memory of:

    Albert (Bert) Green
    Wendy King
    Bill and Charmaine Smallgange
    Glenys Herring
    Tommy Tippit
    Neil Martella

who passed away during the years of struggle described here, and all those affected or displaced by the loss of their town and community, the breakdown of marriages and friendships, and the loss of their families, sense of place and local history.

Community members also wish to acknowledge the late Terry Hahn for his help in documenting community life under corporate skies; the Greens (WA), in particular Dr Christine Sharp, Paul Llewellyn and Giz Watson and their staff for their unwavering support of the community within the corridors of power; John Bradshaw (former

Liberal MLA for Murray-Wellington) for his tireless work in the communities of Yarloop, Wagerup and Waroona. John represented the interests and concerns of residents in Parliament on numerous occasions and did all within his power to lend support.

The community also acknowledges members of the media for their courage, integrity and continued interest in reporting the events unfolding in Yarloop; the executive committee of the Community Alliance for Positive Solutions (Vince Puccio, Merv McDonald, Alex Jovanovich, Terry Cockerham, Lionel Turner, David Puzey and John Harris, and Cam Auxer who was responsible for CAPS' research and communications) for their dedicated pursuit of social justice in the face of enormous odds; Kingsley Dyson for providing IT support; and all those who need to remain anonymous for their help and assistance with the many organisational, technical and legal issues the community needed to become expert in.

INTRODUCTION

This book has been written to provide a space for the marginalised voices in a long-running conflict between residents of the town of Yarloop in Western Australia and their corporate neighbour, Alcoa World Alumina Australia. For years residents have struggled to be heard in their concerns about Alcoa's Wagerup alumina refinery which is located a few kilometres away from the community. We present the local stories about life under corporate skies as we explore the problem of balancing the needs of economic development with people's health, wellbeing and place, and environmental quality.

In fact all parties in the conflict – community, industry and government – get to tell their stories here, though we unabashedly side with the least powerful participants: the community members who are affected by the presence and activities of their corporate neighbour. For these people suffer a lack of resources and access to decision-makers to defend themselves and improve their situation. Because of this power imbalance, we give voice to those unable to be heard in the spheres where politics and economics meet and decisions are made about the fate of regional sustainability.

When we embarked on writing this book, we were both academics working at Edith Cowan University (ECU) in Western Australia, though in different disciplines and on different campuses. We separately began researching the Wagerup controversy several years apart and from quite different starting points. Dyann conducted a two-year action research study in 2002 funded by Alcoa Wagerup at the height of national media reports based on protests from the residents about pollution from the refinery. The aim of the research was to enable dialogue between the company and the community. Dyann and other ECU researchers facilitated a range of forums for identifying and trying to solve the shared problems, with the conflicted nature of company–community relations the main focus of the research.

In 2005, Martin became involved in the conflict in response to concerns by Yarloop residents about the impacts of the Wagerup refinery on their community. The following year he secured independent funding through an ECU research grant to study local experiences of the controversy. Despite the two-year gap between the two projects, there was substantial congruence between the local stories over the period and a worrying persistence of the issues fuelling the conflict.

Material on the public record and the non-restricted documents produced by Dyann's research are used in the book to establish the context and parameters of the issues. Martin's research, based on in-depth interviews with residents, company personnel and government officials, provides rich and contested descriptions of the struggles over what constitutes sustainable development in the region. Overall, during the course of our respective research projects, we spoke to over 500 people between us. This group included twenty-five state government officials and government department staff, twenty Alcoa managers, fifteen independent consultants and advisors and more than 400 community members.

Based on accounts of the complex and difficult relationships between residents, their corporate neighbour and their elected leaders, we try to shed light on the dark side of today's largely unquestioned development agenda, traded internationally under the umbrella term of 'globalisation' [1].

One aspect of this 'dark side' is that not everyone benefits from the current global economic experiment [2]. We show that the benefits usually assumed to result from economic development can largely fail to materialise for local people, in particular those communities at the coalface of development – society's 'disposable humanity' [3, p. 189] living downwind of a multinational company's pursuit of maximum returns.

Yet these same communities can offer the most pertinent responses, based on local knowledge and experience, to the human and environmental dilemmas that an uneven development path brings [4]. For this reason alone it is important to provide their stories and points of view on what matters locally, and examine the implications of these for government policy and corporate conduct.

This book was written during a period of unprecedented economic expansion in Western Australia. Large-scale resource-based developments make WA a principal driver of the nation's economic growth and are a key source of wealth creation [5]. While the state's resources boom receives much attention in terms of the economic benefits and flow-on effects it delivers, little attention is paid to the conflicts that can ensue between industry and local communities. WA is particularly prone to such conflicts due to the coincidence of resource rich areas with human settlements and areas of high biodiversity value [6; 7]. Rising global demand for natural resources renders increasing conflicts between local communities and industry interests almost inevitable.

To this day, the nation's economic advancement has remained a policy priority for state and federal governments alike. Typically, governments foreground benefits such as employment and prosperity, and celebrate win–win outcomes for companies and communities, while being hostile to views critical of their development agenda [8]. Political and corporate elites — those at the helm of government and economic institutions — assume that society supports these development goals and accepts any potential social and environmental costs their pursuit might entail.

The stories in this book, however, foreground a power imbalance between the pursuit of corporate profits and a government's desire to provide an environment conducive to business on the one hand and a community's rights to health and safety on the other. Indeed, these stories make explicit the sustainability dilemma of balancing the pursuit of environmental, social and economic goals, and challenge the assumed acceptance of economic development regardless of cost.

Together, these stories highlight the need to negotiate conflicts arising out of profit, people and place [9]; to care for those affected; and to seek ways in which the adverse impacts of development can be mitigated. While this book cannot solve the local conflict addressed here, it maps and explores its dimensions and dynamics in the hope that by making explicit the different stories on local and regional sustainability, a better understanding of the underlying issues can be obtained. This may serve to inform and possibly avert future conflicts and help improve industry–community relations. It is also our intention that the space provided here may create room for dialogue between the parties, for healing and the mitigation of harm.

The chapters in this book paint a picture of regional sustainability under threat. If the current path is to serve as a blueprint for future development in Western Australia, the resilience of many more local communities and places is likely to be tested and threatened. Thus we argue for a form of partnership building that enables industries, governments and their electorates to meet as equals. Partnerships such as these, based on a common understanding of progress and development, help achieve a shared and sustainable vision, which provides not only for society's material needs but also serves to protect what matters locally: people and place.

Chapter 1 introduces the town and the company involved in the conflict. It also sets out some of the ideas underlying our understanding of the local issues and the analytical lenses we employ to make sense of them – the concepts of sustainability, governance and corporate social responsibility that guide our understanding of the industry–community conflict in Yarloop.

In Chapter 2 we map the history of the Wagerup conflict in greater detail and contextualise the points of contestation that have fuelled it.

In Chapter 3, local people share their experiences of life under corporate skies, providing a detailed account of the ways in which their lives, the community and the environment have been affected by their corporate neighbour.

In Chapter 4 we bring in viewpoints from Wagerup company managers and other company personnel, along with Alcoa's own promotional material about its refinery and community relations. We highlight points of difference as well as commonalities between the dominant corporate discourse on regional sustainability and the marginalised discourses of community members detailed in the preceding chapter.

Chapter 5 introduces the role of government in this conflict, including the viewpoints of some key politicians and government officials. Extracts from relevant public documents further explain the different perspectives on regional sustainability pertaining to Yarloop. In this chapter we address the problematic role of government charged with the responsibility of balancing the needs of industry and community.

In Chapter 6 we map and analyse the conflicting discourses presented in earlier chapters, focusing specifically on the nature of corporate and government conduct and community responses to it. The analysis concentrates on the use of power and knowledge and their respective impacts on local people's identity and sense of place. The contradictions, tensions and silences we discover point to the heart of the sustainability dilemma in regional areas where industries are located in close proximity to towns and high-value ecosystems.

In the final chapter we return to the debates about sustainability, governance and corporate social responsibility, analysing the stories presented in previous chapters in light of these concepts. We focus on the lessons to be learned and call for renewed effort by government in considering regional development, industry–community relations and environmental sustainability from the standpoint of, and inclusive of, an informed, active citizenry. We articulate new rules for a compassionate, socially just and accountable corporate engagement with communities as the basis for long-term corporate viability and as a prerequisite for future sustainability. We highlight the importance of the public asserting and reclaiming its legitimacy as a stakeholder with power. Only through the involvement of communities will it be possible to effectively balance and negotiate social and environmental trade-offs and competing claims for resources – the crux of today's sustainability challenge.

The conflict surrounding the Wagerup alumina refinery is still evolving. To this day, no amount of Alcoan money spent to improve aspects of the company's operations, or government support programs to remedy the situation for the nearby towns, has changed the conflict dynamics or stopped the harm and loss for the local communities.

CHAPTER 1
A SMALL TOWN AND ITS CORPORATE NEIGHBOUR

YARLOOP AND ALCOA

The town of Yarloop, home to approximately 600 residents [1], is located in Western Australia's rich agricultural country on the fertile coastal plain between the Darling Range and the Indian Ocean, about 125 km south of Perth, the state capital. The town was once the cherished home of its residents who saw in it a 'slice of paradise', as some locals recalled:

> *It's just a beautiful little spot ... It's just idyllic. It's a very pretty town and it had everything we wanted. You really couldn't want much more. (Yarloop resident)*

> *Just so different from Perth. Quiet, nice little community. Green, clean, just the sort of place you want to go to get away from Perth and the stress of big business. (Yarloop resident)*

A beautiful little spot (Photo by H. Seiver)

White settlers first arrived in Yarloop in 1849 and their industry heralded Yarloop's long and proud history as a timber town. The town later also became known for its large steam engine works.

A cohesive community: members of the local CWA (Photo by H. Seiver)

Many of the Yarloop residents who feature in this book have a long-standing connection and history with the town. Unsurprisingly, ties such as these help create a strong sense of place and belonging.

Yeah, my grandfather was there and my great-grandfather and great-great-grandfather. Yeah, it goes back a long way. (Yarloop resident)

Dad's family came there in 1910, 1911. Mum was born in Yarloop. They came in 1906, so 100 years of history we have associated with that town. It's very hard to walk away from. (former Yarloop resident)

The town has been much loved by its long-term residents for its strong sense of community.

The social connection, the friendship, the people looking after each other, the way this town was close and worked together. If someone had a problem, there was always someone there to help you out or look out for you. (Yarloop resident)

... even though it wasn't a huge community it was a very strong community and the sort of community where everybody knew everybody; everybody looked after one another ... (Yarloop resident)

Then in 1984 Alcoa's Wagerup alumina refinery began operations, only two kilometres from Yarloop. Alcoa is one of the world's largest producers of aluminium. The US based company oversees operations in more than forty countries and employs close to 130,000 people globally. In Australia, the company trades under the name of Alcoa World Alumina Australia. It operates two smelters and a power station in Victoria, aluminium rolling mills and recycling plants in Victoria and New South Wales, and three alumina refineries and two bauxite mines in Western Australia.

Alcoa: the multinational next door (Photo by John Harris)

In Western Australia, Alcoa's workforce of about 4000 people produces around 7 million tonnes of alumina each year, accounting for 13 per cent of total world demand and resulting in export earnings of around A$2.8 billion [2; 3]. Alcoa prides itself on being a significant contributor to the Western Australian economy as well as a socially and environmentally repsonsible business. The company's achievements in these areas are recognised both nationally and internationally [4; 5].

THE CONFLICT

Since the mid-1990s, residents and Alcoa workers have reported symptoms such as frequent blood noses, headaches and nausea. No causal link has been formally established between the refinery's emissions and people's health, and the matter has been the subject of much local, national and international media coverage, even a Parliamentary Inquiry. The conflict between the community and the corporation has prompted numerous research projects and given rise to sustained local activism. The Standing Committee on Environment and Public Affairs [6] held an inquiry into a wide range of concerns raised by community members. The issues that formed the terms of reference for the inquiry (reported in 2004) are shown below, together with recent statements by residents demonstrating that the issues are not yet resolved.

Public health
And my skin, I get burnt. It's like a radiation thing. You also have bladder problems and it affects your bowel, it affects your moods, it affects your skin, see my skin is horrible. I can't explain; my stomach is always sore after I've been outside and stuff has come on me. (Cookernup resident)

Loss of amenity
There was the pub, there were the shops. There was a butcher, hairdresser, and it was a real community. You could walk around it and the grandchildren would come down, and then you just watched it all disappear. (Yarloop resident)

Social impacts
So this whole town has been fragmented, it's been divided, you call it whatever you want, but it doesn't even have 10 per cent of what we used to have as a community, and we had a very strong community here. (Yarloop resident)

Alcoa's land management strategy
When Alcoa made their buffer zones, they put this dividing line in and then they made two, three classes of people. Area A was looked after ... But the B area people were only offered market value. The C area [people] weren't offered anything. So there was infighting, the town people on one side of the fence were fighting the other side of the fence. So that's the beginning of all the changes. (Yarloop resident)

Responses to community concerns by Alcoa and successive state governments over the last decade have only served to increase the intensity of the distrust. Initiatives taken by the company were met with suspicion by locals, for they felt that:

... [The company] tried to cover up exactly what was happening; they tried to say that it was safe; that there was nothing to worry about. (Yarloop resident)

Suspicions were further heightened by the way in which the company was seen to engage with the community and to respond to its concerns:

> *They came in and they dish[ed] out promises and promises, but they're lies and lies and lies. They end up doing what they want. They're a bully. They kick the little guys when they're down, and there's no compassion at all there. (Yarloop resident)*

Many residents felt betrayed by their elected leaders in state government who they thought:

> *... were supposed to be watchdogs. They're supposed to be protecting the basic rights of their citizens. And it's the UN that said ... that it's a basic human right to have a clean environment to live and work in. And I think the government is there too, yes, to see that our society stays afloat, that its economy should be looked after, granted. Development needs to be sustainable. They need not be short-sighted. But, certainly, the rights of their taxpayers and their residents need to be protected. The environment needs to be protected. And if you look at what's happened down here, they have failed miserably. (Yarloop resident)*

Residents took exception to the fact that their concerns appeared to be downplayed by the authorities. It seemed that 'the corporate dollar was outweighing the health of the community and the environment' (Yarloop resident).

Events are still unfolding in Yarloop. In 2006 the state government approved a major expansion at the Wagerup refinery – despite community concerns and reservations voiced by the WA Health Department [7] as well as independent medical experts [8]. The decision was announced by the former Minister for the Environment, Mark McGowan:

> *I have decided to grant environmental approval to the expansion of the refinery subject to 42 conditions*

dealing with project design, emissions, noise, dust, water management and residue disposal. The conditions I am proposing are more stringent than those recommended by the Environmental Protection Authority (EPA) and will make the refinery one of the most tightly regulated in the world. [9]

Scores of local submissions to the EPA against the expansion referred to loss of social amenity, harm from fear and the effects of pollution, as well as concern for devalued assets and loss of family and friends from the area. One example is provided here in full to convey the emotional upheaval experienced by local residents.

Local Residents' Submission Against Alcoa Wagerup's Expansion
23 July 2005

We wish to let our views and concerns be known about Alcoa's efforts to get an expansion at its Wagerup refinery. This is totally unacceptable to us and threatens our sense of safety and wellbeing after what has already been years of adverse social and health impacts from the refinery.

We lived in the northern fringe of Yarloop happily for many years until Alcoa installed the liquor burner in 1996. Since that time Kay has suffered quite debilitating health effects from direct exposure to airborne pollution from the refinery. Alcoa staff have even witnessed her vomiting and her distress when responding to our complaints. We have kept a detailed logbook of all the times we have lodged a complaint with Alcoa, each time corresponding with personal suffering on my part in witnessing my wife's failing health. There was a period when I was really concerned I was going to lose Kay due to the deterioration in her health. She became trapped in the house, which is no way to live.

Neither of us wanted to move from our home and close contact with long-term friends in Yarloop. But as Kay was so unable to lead a normal life we had no choice than to eventually take up Alcoa's offer to sell to them. We bitterly resent having had to do this and haven't yet recovered from the loss of our home in Yarloop. We are now living in Cookernup and, with all this talk of an expansion at Wagerup, are experiencing a heightened fear that we will now be impacted here as well.

In the last month I have had several nosebleeds which is very alarming as I haven't had any since leaving Yarloop. One of the nosebleeds occurred when I was visiting a friend in Yarloop. We are worried that it will continue and get worse for us and it doesn't make sense that Alcoa says the expansion will not result in an increase in noise, air pollution and the like. As it currently is, it's a problem so we can't in good conscience believe them that it won't be in the future.

Not only have we lost many of our friends who felt they had to leave for their own safety and to protect their financial interests but we still find many of our conversations in the community dominated by talk of Alcoa. This industry is impacting too much on our everyday lives and is much too determined to have its own way at our expense. There is already plenty of evidence that Alcoa and the government are aware of the social impacts of the refinery operations on these communities. What seems to be happening is a quick patch-up by throwing some money to some community groups and thinking this fixes everything. It is much too soon to be expecting those of us who have been so seriously threatened by Alcoa to be presuming an expansion is acceptable. People and communities need to feel safe and able to survive with the current levels of production before an expansion is even considered. That there is an increase of large proportions in their production already happening leaves us disturbed. How is this happening even before the current application is heard?

We are also alarmed at the West's report of a spill at Wagerup this week. This is no surprise to us and we suspect the delay in them reporting their claim that it wasn't, according to their judgement, a risk is political, as the last thing they want at the moment is such adverse public attention.

We are concerned that the little people who are most impacted and least able to run weekly advertising programs about our experiences (compared with Alcoa in recent months, promoting their credentials and how good the expansion will be for us all) will not be heard. Alongside this we have no confidence that Alcoa knows how to be good neighbours to those of us who are badly impacted.

It can't be left to them to say what we need and what the social initiatives they can provide are. They have yet to fix the problem and yet are pushing for an expansion for purely economic reasons. This feels to us like a blatant disregard for recent history and the continuing controversy about the social impact in this area. We are just one example of how the situation is still affecting local folks.

Despite widespread opposition and the many public submissions echoing the local concerns captured above, the expansion was approved in September 2006. The company welcomed the approval by the state government, speaking of a win-win outcome for both Alcoa and regional communities.

Green light for alumina refinery expansion

Alcoa World Alumina Australia Managing Director, Wayne Osborn, said the planned expansion of Alcoa's Wagerup alumina refinery will provide major social and economic benefits for Western Australia. Speaking after the WA government today gave formal environmental approval for the project to proceed, Mr Osborn said the expansion would create over 1500 construction jobs as well as 3000 direct and indirect jobs, including 260 new permanent Alcoa jobs.

'Alcoa has committed to implementing the expansion with no increase in noise, dust or odour impacts, and extensive scientific investigations have shown both the existing and expanded refinery are safe for our employees and neighbouring communities,' Mr Osborn said. 'Regional businesses and communities would also reap the benefits of significantly increased local spending. The Wagerup refinery already spends over A$40 million a year with businesses in the local area and this will rise sharply.' [10]

GLOBAL AND REGIONAL BENEFITS — THE LOCAL COSTS

The stories presented in this book convey competing perceptions – by town residents, company personnel and government spokespeople – of gains and losses as well as acceptable risk. Government approval of Alcoa's expansion dovetails with its agenda to drive economic growth in the state – hardly controversial since economic development is assumed to improve the human lot, a notion which, to this day, has largely gone unchallenged [11]. Undeniably, Australia is enjoying record levels in household income and historically low levels of unemployment, both attributable to strong economic growth [12]. Unsurprisingly therefore, both Alcoa and the state government – each operating under the banner of a self-professed sustainability framework [13; 14] – emphasise the economic benefits of the refinery and its expansion. Within their respective frameworks, economic growth is largely seen uncritically as a catalyst for regional sustainability.

According to Alcoa's former CEO, Wayne Osborn, the 'Wagerup expansion would ... provide over A$11 million a year in extra state government revenue' and result in an increase in community funding 'to support local infrastructure and projects in the Harvey and Waroona Shires.' In fact, Osborn says, 'Alcoa's A$400,000 a year contribution would almost double under the expansion' [10]. Alcoa portrays itself as a company that is helping build a sustainable future [15]: 'Alcoa is committed to contributing to positive sustainable outcomes for the communities of the south west region. During the life of the Wagerup refinery, Alcoa has helped establish a long-term sustainable future for Waroona, Yarloop, Hamel, Harvey and the region through its contribution to:
- local infrastructure and services
- local community organisations
- local and regional development
- regional and state infrastructure, and
- community based education and training' [16, p.13].

The state government supported the refinery's expansion with the aim of pursuing 'jobs and opportunities for Western Australians, but not at any cost.' The government claimed that 'the wellbeing of people in Yarloop and surrounds [had] been central to [their] thinking' [17]. Moreover, 'the assessment of the Wagerup proposal was [said to have been] the most complex undertaken by the [Environmental Protection] Authority (EPA),' because of 'the plant's history of health-related complaints.' The conditions placed on the expansion were said to be the 'most stringent conditions the EPA has recommended for any industrial or mining project in Western Australia' [18].

In the shadow of promises of economic gain and environmental protection however are social and environmental indicators that highlight the side effects of uncontrolled development [19; 20]. Local communities, like canaries in the coal mines, are the first to detect perceptible risks to human wellbeing. In Yarloop, residents have detected some of the impacts of development in their community. Many recognise the economic benefits of industrialisation; they are not simply anti-development. Based on their experiences with the Wagerup refinery, however, they are concerned about development that brings regional and global benefits at local costs.

There are economic benefits to the whole state but I think the local people shouldn't bear the brunt of the progress. (Cookernup resident)

We're not here to shut Alcoa down. We're here to make them accountable. (Yarloop resident)

Some residents feel they would be paying for the company's proposed expansion, and the relationship between government and industry is questioned mistrustfully now by many. The photograph below depicts one resident's anger in the form of a protest he took to many high traffic spots during the peak of the controversy.

Not only is the refinery believed to be having an adverse impact on the community's health and wellbeing, other aspects of Alcoa's operations are also seen to be directly threatening the region's sustainability. By its own admission Alcoa is a major emitter of greenhouse gases in Australia [21] and one of the heaviest users of energy and fresh water in WA [2; 22; 23]. The sustainability of clearing native jarrah forest for the mining of bauxite, a

Silent protest (Photo by V. Webb)

non-renewable resource, is challenged by conservation groups who for many years have also criticised mining companies' land rehabilitation practices and warned of the spread of disease such as dieback (*Phytophthora*) through mining operations [24; 25].

Despite a growing awareness of the importance of local solutions to global sustainability [26; 27], residents' voices rarely register in debates about current issues and future sustainability. Publicly listed companies, often in partnership with government, tend to determine the eventual balance between economic, social and environmental concerns [28]. This government–industry relationship has been called into question by the Yarloop and Districts Concerned Residents Committee, which formed in 2001 in response to the impacts Alcoa's Wagerup refinery was seen to be having on the community.

Excerpt of submission against Alcoa's expansion

It has been our experience that despite what the Department of Environment (DoE) say to us and/ or agree to, the DoE continually demonstrate a high level of bias towards Alcoa. There have been numerous examples of this behaviour, of which we

> *have evidence and we believe this department should be independently and publicly investigated. [29]*

The people who claim to be affected by Alcoa are largely 'out of sight and out of mind' due to their rural location and the very small population of the towns. The plight of the rural population may not resonate with the population of the urban centres, which holds much of the political muscle. An even greater distance exists between local realities and the decision-making processes at the corporate boardroom level. Alcoa promotes itself as a company that cares to the extent that 'people in head office who have never even visited Yarloop' are very much affected by what has been happening and 'feel very much for the people of Yarloop and the impacts on them' [30]. Locally, however, there is a sense that 'they're tens of thousands of kilometres away, who cares what happens here? Any decisions made in America are not for our benefit' (Yarloop resident).

The media has assisted greatly in giving voice to people's concerns, helping to bring the Yarloop issue onto the political agenda. The following story appeared in the *Sunday Times* in 2006.

Alcoa expands despite toxins

> *Mr Royce, a local farmer, believes toxic emissions from the alumina refinery caused his wife, Jill, to develop cancer in 1996 – the same year the controversial liquor burner fired up. She died in 2002. But Mr Royce didn't join the campaign by locals opposing the plant's expansion because he thought it would be futile. 'You're never going to have a government worry about a small group of local people on one side and a billion-dollar industry on the other,' he said. 'Also, the people who live in Perth don't give a damn. The voting power of this state could not care less.' [31]*

It is easy for political and commercial decision-makers to frame concerned residents as 'noisy' activists, stirring up anti-company sentiment. However, sidelining their concerns in this way means the underlying issues do not get treated seriously. It has taken a Parliamentary Inquiry to begin to redress this invisibility. Still, the capacity of local people to influence Alcoa's development decisions and to defend the viability of their town is severely limited by the lack of adequate participatory and consultative forums [32]. In its defence, Alcoa says:

> *Western Australia's most comprehensive community consultation process was developed and implemented early for this [Expansion] Project Proposal ... over 60 community working groups meetings, over approximately eight months were held totalling more than 200 cumulative hours of consultation. [16, p.11]*

Impressive though this may sound in terms of hours, this consultation process appeared to be driven by Alcoa's expansionist agenda and the company continued to maintain that 'the threat of serious illness from the refinery is negligible ... with no long-term health risk' [33]. Key activist groups avoided the table because they saw the process as stacked against informed and critical perspectives. The company's repeated dismissal of community concerns served to harden their views, as shown in the submission from one group below.

Public submission against refinery expansion

> *There has been absolutely no recognition, discussions or proposed methods to mitigate, control or manage amenity issues caused by dust, noise or odour from the mine site, RDAs [Residue Drying Areas – the waste deposits of the red mud which is the by-product of alumina production, often referred to as mudlakes] or Refinery from the existing refinery operations at*

Wagerup. Nor has there been any discussions or even consideration of amenity for the expansion.

We believe there is an ongoing failure of the 'system' that needs to be addressed prior to consideration of this expansion.

If these shortcomings are not identified and addressed, no amount of reports and studies, nor conditions and commitments will be of any benefit to those affected. The fact is that all of the above is of no use currently. [29]

TALKING ABOUT SUSTAINABILITY

Only with a balance between profit, place and people can the goals of sustainability be reached. While all parties in the Yarloop conflict speak of sustainability, they employ different understandings of the concept. Throughout this book we explore some of these different understandings as we unearth ways in which the sustainability of a local community is affected by decisions about regional sustainability made by industry and government.

In broad terms, sustainability aims at the harmonisation of environmental, social and economic goals for the benefit of human and environmental health, economic wealth and equity [34; 35; 36; 37]. In corporate terms, sustainability translates into the economic balance between people, place and profits [38]. However it is defined, the environmental, social, political and economic dimensions of sustainability cannot be treated as separate from each other because they are interdependent and affected by the actions of people across these domains [39]. Neither the economy nor society operates in a vacuum.

Society as a whole can be assumed to share a common goal: the future sustainability of humanity and the planet. Regardless of the extent to which behaviour coincides with these ideals, over the last decade, sustainability has been increasingly accepted as the guiding development goal [40]. Questions of 'how' and 'by whom' however remain contested. Today's sustainability debate is driven largely by governments and industry, and, at the level of theory, by university researchers such as ourselves. Political and economic forces tend to determine the eventual balance between economic, social and environmental concerns [41; 42; 43], and the search for the tools to shift societies onto more sustainable pathways is thus skewed towards those who are relatively privileged.

Decisions about sustainability not only tend to exclude the public, they also frequently silence [44] and suppress [45] any form of diversion from the dominant econo-political understanding. This absence of public input into policy debates precludes open and transparent discussion of the trade-offs that are made between peoples' rights, the environment and corporate profits.

Thankfully, the dominance of politics and profit in debates about social and environmental goals has come under attack in recent years and the appropriateness of economically driven assumptions underlying public policy making and long-term sustainability of today's neoliberal growth agenda [46] is increasingly questioned. In particular, there have been renewed calls for active citizenship in processes of policy formulation [47; 48; 49; 50; 51].

This book is posited on the view that 'debates over sustainable development require equal and adequate representation of the communities affected' by development [52]. Juxtaposing the many different stories addressing questions of regional sustainability exposes the tensions between the various understandings of sustainability. We show in the end that the conflict described in this book is a struggle between competing and, to some extent, irreconcilable interests.

GOVERNANCE AND CORPORATE SOCIAL RESPONSIBILITY

Sustainability relates to questions about the role of government in setting development goals and balancing those goals with social and environmental needs. It also relates to how industry engages with local communities and manages their role in regional development. These twin concepts of governance and corporate social responsibility (CSR) provide useful lenses through which to observe and analyse the complex interrelations between industries, governments and communities relating to the sustainability agenda [53]. They advocate the responsible and ethical conduct of political and corporate leaders for the protection of public, and increasingly environmental, interests.

In these terms, the benevolent and prudent use of political and corporate powers are assumed to safeguard people and the environment from the uncontrolled exercise of such powers. Social stability and institutional robustness are attributed to good governance [54; 55]. Sound CSR practices are believed to ensure environmental and social acceptability as well as corporate profitability. Dominant strands of theory on governance and CSR even suggest that public and corporate interests correlate and overlap [53; 56], leading to the conclusion that the goals of governance and business can be merged since the creation of an environment conducive to business is understood to be in the public's best interest [57].

The Yarloop conflict challenges all these assumptions.

When dissenting stories of local people are dismissed by corporate interests and government, the limitations of governance and CSR are revealed. Here, the sustainability agenda becomes mired in the dominance of profit maximisation and the political pursuit of economic growth. Ideals of corporate responsiveness to the community and governmental regulation of development in response to social and environmental policy needs are celebrated more in their breach than in action. Dissent is marginalised or ignored at best, attacked and discredited at worst. This refutation of society's expressed needs in turn contributes to social, economic and political imbalances that threaten the entire sustainability enterprise.

We use these concepts of sustainability, governance and CSR to guide our understanding of the conflict in Yarloop as we argue for a socially just and compassionate approach [58], beyond economics and profits. To this end we foreground and re-legitimise the marginalised local voices within a much-needed broader and deeper sustainability debate.

CHAPTER 2
THE PATHOLOGY OF INDUSTRY-COMMUNITY RELATIONS

In this chapter we present a chronology of the key events of the Wagerup conflict between 1996 and 2007. In it we emphasise local perspectives to illustrate how past events have affected, and continue to affect, community dynamics and how they have shaped local views on the refinery.

Geographically, the refinery is located just to the north of the Shire of Harvey, between the towns of Waroona and Harvey, along the South-West Highway approximately 125 km south of Perth. Small towns and localities in the vicinity of the refinery include Yarloop, Hamel and Cookernup. Yarloop is the closest town to the refinery with its northern boundary just two kilometres from the plant. Wagerup, the site of the refinery, used to be a small town; it was gradually displaced by the growing refinery and as such no longer exists.

The Wagerup refinery was controversial from its inception. Bauxite mining and alumina refining had triggered waves of environmental protests in the early 1960s, largely organised by Perth-based environmental groups such as the Campaign to Save Native Forests (WA) and the South-West Forests Defence Foundation [1; 2]. According to a conservationist involved in the protest at the time, there were concerns even then that:

> ... *bauxite mining was already removing a significant area of jarrah forests from Perth's drinking water catchments ... and that the risks of forest regeneration being unsuccessful were too great to increase the scale of bauxite mining in the Darling Range.*

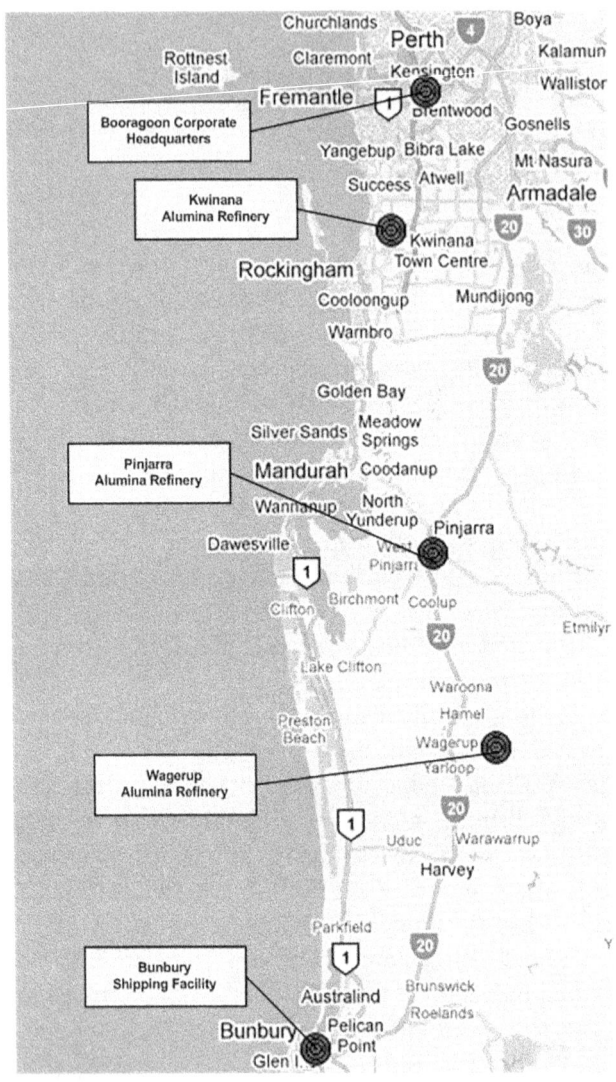

Map of Alcoa's operations [3]

> *Bauxite mining and particularly the construction of this new refinery ... would greatly increase the rate of clearing.*

Conservationists at the time challenged the viability of bauxite mining and alumina refining in terms of poor environmental outcomes and meagre economic returns, given the loss of native forest and the industrial use of scarce freshwater resources.

> *I saw it as a compromising of our state ecosystems for actually very little return for Western Australia ... we were alerted to the fact that the grade of bauxite that was being taken out of the jarrah forests was some of the lowest in the world so the reason that they were able to do that was that they got very favourable arrangements in terms of access to water and the amount of royalties they paid for the bauxite, and of course how much they ... had to actually put into restoring the areas that they had mined. So it seemed to be crazy on a whole range of levels, especially as it was a multinational company that wasn't even necessarily returning much profit to WA, other than the jobs that it actually created. (former environmental activist)*

These concerns were echoed years later in a study of Australia's aluminium industry which found the industry to incur a net loss to the Australian taxpayers when all (largely secretive) [4] government subsidies (e.g., water and electricity) as well as associated social and environmental costs were considered [5].

There was also unease among locals regarding the secrecy surrounding the arrival of the alumina industry in the area when land purchases for the Wagerup refinery were conducted surreptitiously in the mid-1970s [6; 7]. According to residents and media reports, farming land was bought above market price under the guise of a sunflower operation, presumably so as not to raise suspicion or trigger opposition to the industry on the grounds that industrial land use would be incompatible with local dairy farms [8] – as had happened with other large industrial ventures in the state [6]. Other rumours at the time suggested the land might have been bought for a beach sands project or a dairy combine [7]. As some locals recall:

> *... Put it this way ... when they first started buying land here in '78, I think it was, they come in under the guise of a plan for sunflower farming, right? That's how it'd come in ... That's their way of being sneaky, and their tactics haven't changed, they've only got worse. (Yarloop resident)*

> *Their phoney [doings] started, I think it was going back to around '79, '80. They were buying land, and they came under false pretences then! It's not Alcoa buying land. It was a real estate agent in Perth that was buying land ... They wanted areas there where there was an abundance of water. First they wanted to grow sunflowers; then they wanted to grow some hemp or something like that, I can't remember ... They're buying all this land because people did sell it to begin with. They were offering big money to start off with. There was a handful of farmers that sold but then there was another handful that were digging in [saying] 'I don't want to sell.' They didn't want to grow sunflowers. They wanted the whole bloody area; they wanted the whole block. Lies from the beginning, and they've never stopped since. (local farmer)*

Following the public announcement of Alcoa's arrival in Wagerup, some residents expressed concern about the presence of an alumina refinery right at their doorstep [9; 8]. This led to the formation of unusual alliances between local people and conservationists, groups that traditionally faced each other from opposite ends of the socio-political spectrum. One local farmer shares his memories:

> *I actually supported the bloody greenies. I didn't want the bastards here; well, not to build the refinery where they did because it was prime bloody dairy land. Because once we found out ... [that] at the back of the Cookernup – the Wagerup hall – was all going to be mud lakes ... I thought, 'Shit, that's bloody close to town.' But any rate, we got steamrolled.*

Overall, however, the establishment of the industry was welcomed for the prospect of local employment and income amidst a decline in traditional industries in the area such as timber milling and dairy farming [10].

> *Alcoa opened and then everybody, I for one, thought it was the best thing since sliced bread; we all did. We all had jobs up there ... When Alcoa came we thought, 'Oh beauty, we've all got jobs in the future.'*

Western Australian premier at the time, Sir Charles Court, and members of the Waroona Shire were especially supportive of the refinery construction, believing it to be of benefit for the district because of the employment it would generate [10]. Indeed, the Court government was reported to have been 'exploring every possible avenue to bring the ... project to fruition' [11, p. 8], calling its eventual establishment 'a great moment for WA and the nation' and 'a major lift-off in industrial development we [the government] ha[d] been working towards' [11, p. 1].

In the face of growing public protests the Court government passed legislation to advance the development of the refinery independent of input from the state's

Environmental Protection Authority (EPA). The government also strengthened police powers to prosecute protesters in an attempt to minimise impediments to the construction of the refinery by way of changing the *Police Act* [12]. We return to this demonstrative support for the industry by the state government in later chapters.

Approval for construction was granted under a *State Agreement Act* [13] in 1978 following approval by the state's EPA [14], despite concerns about potential problems with the siting of the plant, related to topography and prevailing weather conditions.

LOCATION, LOCATION, LOCATION

The location for the proposed refinery was near the foot of the Darling Range, a low-lying escarpment that extends north–south to the east of the Swan Coastal Plain. Questions were raised about the proximity of the refinery to the escarpment and local communities. There was concern, which Alcoa acknowledged in its 1978 Environmental Review and Management Plan [15, pp. 325–8], that emissions could be trapped during inversion weather and concentrate in the atmosphere, leading to the exposure of nearby communities to concentrated refinery emissions. As indicated by an independent consultant:

> *Having the township of Yarloop and the refinery location close to the scarp probably meant a less dispersive air movement would occur; that there would actually be a concentrated plume of the emissions go through the town with relatively little dilution.*

Concerns with the location were also noted by the EPA at the time [14], who hinted at possible air pollution problems because of the proposed site's proximity to the Darling Range and the characteristics of the airshed around the area. Yet, no remedies were being proposed. The EPA

The Wagerup refinery at the foot of the Darling Range
(Photo by Community Alliance for Positive Solutions Inc.)

had prevented the construction of an alumina refinery in Perth's Swan Valley only a few years earlier, which may explain its reluctance to block yet another development proposal under the very growth-orientated Court government of the day.

In hindsight, the siting of the plant near the escarpment is considered a mistake, as noted by Hon. Bruce Donaldson in the Western Australian Parliament [16].

There were inherent problems upfront. One of the other problems was the site. There are [catabolic] winds off the Darling Escarpment. Members who have lived close to that escarpment will be aware that the easterly wind that roars down, and the inversion layers that occur at certain times of the year, cause problems. If Alcoa were to start all over again, the refinery would probably not be built at its current site. That is a fact of life; but it is there.

For Alcoa, this was a matter of poor planning on the part of the government. The lack of an 'adequate buffer' and 'the close proximity of the refinery to ... residential landholdings' was recognised by former Alcoa CEO Wayne Osborn as 'a root cause of the problems at Wagerup' because it resulted in 'land uses that rested uncomfortably with refinery operations.' Yet the absence of a 'coherent formal land use framework' was seen by the company as something it had 'no control' over, it being the responsibility of the 'State and local planning authorities' [17, p. 467].

The Wagerup refinery began operations in 1984 with an annual production volume of 470,000 metric tonnes of alumina, which was increased gradually to an output of around 2.19 million metric tonnes by 1997 [18]. In the first ten years of operation Alcoa and adjacent communities maintained good relations. The company was seen as an economic lifeline for the region, a good place to work and a good neighbour.

Alcoa did have a good name ... Alcoa did seem to be doing the right thing. They did treat their employees well, they were treating the community well.

Industry–community relations started to change in 1996 when Alcoa commissioned a liquor burner facility on the Wagerup site. It had installed the same technology at the company's alumina refinery in Kwinana eight years earlier.

Alumina is produced using the Bayer process, which involves the use of caustic liquor as a means of dissolving alumina out of the bauxite ore. Western Australian bauxite, while abundant contains significant amounts of humic acids [19], organic compounds that decompose in the caustic liquor stream and contaminate the alumina. Liquor burning removes the organic compounds in the liquor stream, enhancing both the quality and quantity of alumina yields as well as resulting in energy and broader cost savings. However, liquor burning technology is controversial due to the associated release of volatile organic compounds (VOCs) [20] known to be hazardous to human health [21].

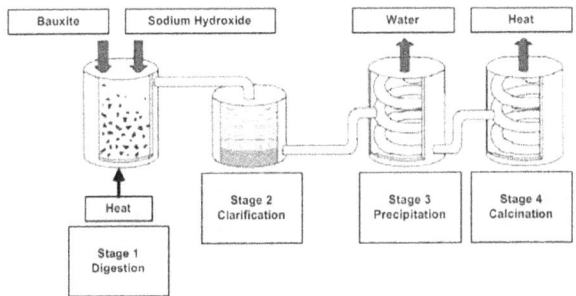

The Bayer process

Aluminium does not occur as a metal, but must first be refined from bauxite in its oxide form. The oxide is called alumina or aluminium oxide, a white granular material that is a little less coarse than table salt. Alumina is produced in the Bayer refining process, which is used by alumina refineries worldwide. The process involves four steps: digestion, clarification, precipitation and calcination [22].

Following the installation of the liquor burner at the Wagerup refinery Alcoa faced a sharp increase in health complaints by workers at the plant. These were similar to the health problems reported by workers at the Kwinana refinery years earlier [23; 24]. Community complaints about emissions and health impacts also increased markedly. These should not have come as a surprise to Alcoa, since the potential for health repercussions caused by the liquor burner was known to the company as early as 1990. A company internal document [25] lays out the concern of technical company staff after a number of aromatic hydrocarbons (including toluene and benzene) had been identified in samples from the liquor burning stacks [26; 20]:

> *Although their concentration is unknown at this stage, we should do some 'homework' on how we communicate this information since many of the compounds are known carcinogens – some of them potent carcinogens.*

Not only did Alcoa know about the possible health impacts, it sought advice on how to respond to potential health complaints from workers and community members. A commissioned report to the company warned [27; 28]:

> *Any unusual diseases (lymphomas, cancers) and possibly the more common ones (asthma, bronchitis) may have to be defended in court, again likely to be costly in time, resources and public relations ... Cancer is a major concern to all communities ... This is best managed by legitimising the dread.*

In response to growing community agitation and the growing public profile of the emission problems linked to the liquor burner at Wagerup, Alcoa shut down the burner in November 1997, installed emissions control equipment, and recommissioned it in May 1998. Between 1999 and 2001, Alcoa continued to install more pollution control equipment with the aim to further reduce VOC emissions. Alcoa subsequently claimed to have been successful in reducing VOC emissions by over 90 per cent [20], but community complaints were ongoing. In fact, agitation intensified in 2001 when Alcoa's application to increase its annual alumina production at Wagerup from 2.2 to 2.35 million tonnes succeeded, despite unresolved health issues.

HEALTH CONCERNS

Between 1996 and 2001 various studies were conducted on the health effects of the Wagerup refinery. However, no toxicological link could be established between refinery emissions and health symptoms in the community [29; 30], nor were any anomalies identified in terms of higher cancer rates in the area [31]. Nonetheless, research with Alcoa workers revealed strong perceptions among participants that the liquor burner triggered certain health effects [32], and a study commissioned by the Department of Health indentified commonalities between the symptoms experienced by individuals who felt affected by the refinery [33].

In response to these findings the state government formed the Wagerup Medical Practitioners Forum to discuss and investigate health problems in Alcoa workers and community members. The forum concluded that there was evidence of a medical problem at Wagerup and described it as an 'association between health issues and Alcoa's refinery' [34]. This was the first time that health problems were acknowledged publicly. The forum supported measures to limit exposure via further emission reductions and the creation of a buffer zone, and called for improved clinical care for individuals who were believed to be suffering from multiple chemical sensitivity (MCS) [34], an allergic reaction to chemicals introduced into the environment. The forum's reference to MCS was also supported by Dr Mark Cullen from the USA, Alcoa's Chief Medical Officer, who advised the company to accept full responsibility for complete and effective remediation of environmental problems at Wagerup [35]. At the same time however, Dr Cullen denied there was any threat of serious illness from the refinery [36] and deemed cases of MCS to be related to psychosomatic, not physical, causes [37].

The subsequent establishment of the Yarloop Community Clinic by the government in October 2002, staffed by a specially trained occupational health nurse, was meant to provide community members with a point of contact to record health effects from specific 'odour' events and to case-manage community members who were experiencing health problems. After six months of operation, a report identified commonly presented symptoms such as headache, fatigue and muscle cramps, as well as sneezing and coughing [38], but did not indicate a cause of the symptoms, and a systematic health study was not conducted. The clinic ceased operation in late 2003. Another health report was published in 2004 [39], and this did not show any statistically significant health anomalies among refinery workers either.

More air monitoring in 2008 revealed possible underestimates in Alcoa's emission inventory: refinery emissions containing up to 260 chemicals were found to be lingering close to the ground for up to 18 hours within 7 km of the refinery [40]. This led to a tightening of Alcoa's licensing conditions and more stringent air-quality controls [41]; community concerns appeared to be vindicated [42]. A health survey conducted in the Wagerup area the same year revealed elevated rates of symptoms potentially related to chemical exposure of residents, and elevated rates of cancer in Cookernup [43].

NOISE BUFFER AND LAND MANAGEMENT

The management of noise at the Wagerup site was another issue. The company applied for approval to increase its allowable noise levels to 47 dBA under Regulation 17 of the Western Australian Environmental Protection (Noise) Regulations 1997, and while the approval was still pending, also sought to implement a noise buffer zone of its own. In late 2001 Alcoa released a land management plan (right) that divided the town of Yarloop into two discrete land management areas.

The company proposed to purchase homes in Area A, offering residents 135 per cent of the unaffected property value plus a $7000 relocation allowance. For residents outside Area A (later known as Area B), Alcoa merely offered to underwrite property values for a period of five years [44]. The proposal was rejected by close to 75 per cent of residents responding to the survey.

Alcoa saw its land management strategy as consistent with the recommendations of the Wagerup Medical Practitioners Forum, which favoured the creation of a buffer zone. The proposed strategy was meant to [45]:
- protect property values
- support the integral nature and quality of the community and encourage people to stay
- make it easy for those wishing to leave to sell their properties.

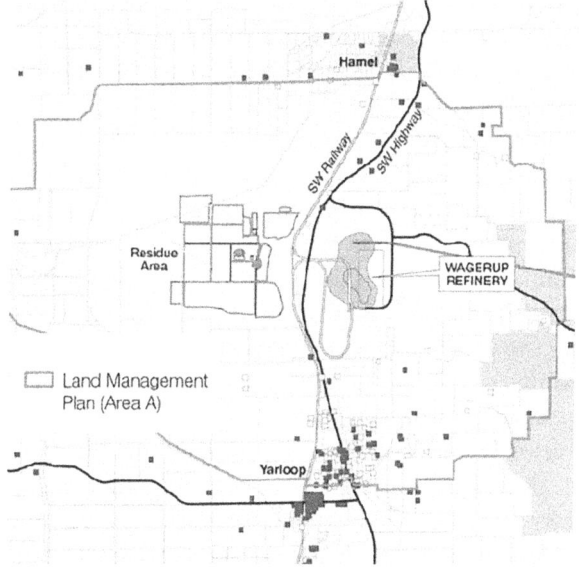

Alcoa's land management plan [1]

According to Alcoa, the boundaries for Area A were determined by noise contours. Residents in Area A were exposed to night-time noise levels from the refinery in excess of the allowable 35 dBA. Area A was also the source of most 'odour' complaints, and was the area planned for future expansion of the bauxite residue area [46]. Alcoa implemented a revised land management strategy in 2002 [45] in which it agreed to guarantee the purchase of properties in Area B at unaffected market value for a period of five years. Due to community pressure and the recommendations of the Parliamentary Inquiry [20], the company later agreed to extend this guarantee for the lifetime of the refinery [47].

It needs noting that Alcoa's land management plan involved the creation of an informal buffer only; it lacked formal status in planning schemes or legislation and did not offer the benefits of a government buffer to separate the refinery from the residents. Such a buffer would have included all of Yarloop and the nearby town of Hamel, meaning all of areas A and B [48].

The creation of the informal buffer zone resulted in the physical partitioning of the town of Yarloop and created sharp social divisions in the community. The differing treatment of residents in areas A and B changed the dynamics of the Wagerup conflict, for it was no longer being driven exclusively by fears of health impacts. Now there were also equity and financial concerns.

While the informal buffer was seen by residents as an in-principle admission that something was wrong with the refinery, its creation on the grounds of noise exposure was also never fully believed by community members. They saw the boundary selection as economically motivated: Area A excluded the more densely populated part of Yarloop, which they presumed was to prevent the high cost associated with property purchases. In any case, noise levels above 35 dBA were believed to extend far beyond Area A. Also, there was doubt whether the company would be able to operate even within the higher noise limit [49].

In 2002 community agitation over the land management strategy prompted Alcoa to form a partnership with staff from the Centre for Social Research at Edith Cowan University (ECU) in Bunbury. This partnership was designed to facilitate a process for resolving conflict over issues of land management [45] and to build constructive and collaborative relationships between Alcoa and property owners in Yarloop and Hamel. Much of this book is informed by this study [50]. Through a series of open, facilitated meetings with company representatives and community members, an attempt was made to establish common ground between the parties [51]. Yet, towards the end of 2003 the project stalled. The community believed that the company refused to meet their demands and publish the findings of ECU research staff, and the project was aborted when ECU researchers resigned [52].

THE STANDING COMMITTEE INQUIRY

The ongoing controversy prompted an inquiry into the Wagerup conflict by the Standing Committee on Environment and Public Affairs in 2001. The committee sought primarily to investigate the operations and impacts of the refinery and to determine the adequacy of responses by government departments and agencies to the problems. After three years of inquiry and over 70 witness statements taken at more than 20 hearings, the committee handed down 22 recommendations, a selection of which is presented below [20, pp. 4–7]:

> *Recommendation 1: The Committee recommends that Government agencies and regulatory authorities should use the term 'emissions' rather than 'odours' to describe general emissions from the refining process.*

> *Recommendation 10: The Committee recommends that the Department of Health should ensure access to appropriate medical expertise and diagnostic health and support services for people with multiple chemical sensitivities and other chemical injuries.*
>
> *Recommendation 16: The Committee recommends that the Department of Environment should assess licensed industrial premises in Western Australia to determine the appropriateness, in each case, of requiring continuous emissions monitoring.*
>
> *Recommendation 18: The Committee recommends that the State Government take critical note of current breaches of the existing noise limits for Alcoa's Wagerup refinery in its consideration of the proposal to increase production from the refinery.*
>
> *Recommendation 22: The Committee recommends that the Department of Health, as a matter of priority, derive a hazard index for locations near to Wagerup in order to assess the health risks caused by the cumulative impact of the very high number of chemicals mixed together in the emissions from Alcoa's refinery at Wagerup.*

The committee wanted more research to be undertaken into the health issues surrounding the facility and for greater transparency of the company's operations. It found Alcoa [20, p. 370]:

> *Failed to adequately recognise and respond to the complaints it received from workers and the local community.*

The committee also criticised Alcoa and the Departments for Health and the Environment for failing to offer an unequivocal and comprehensive response to what was described as [20, p. 370]:

A range of extremely serious and complex issues at Alcoa's refinery at Wagerup from 1996 to 2001.

The committee vindicated many of the concerns raised at the community level, yet many problems identified during the inquiry persist to this day as government and its departments and agencies failed to implement the recommendations handed down by the Standing Committee. This is addressed when we turn to the role of government in Chapter 5.

ATTEMPTS AT COMMUNITY CONSULTATION

As the conflict evolved, Alcoa recognised the need for dialogue with local community members and formed a

Point of view of a long-term resident

dedicated Community Consultative Network (CCN) to enable community feedback and input [53].

It was also acknowledged that the Wagerup conflict needed to be addressed at the level of government. The Welker Review [54], which reviewed environmental licensing strategies on behalf of what was then the Department of Environmental Protection (DEP), identified community dissatisfaction with the department's licensing process and outcomes, citing this as a reason for the loss of confidence in the process. The final report highlighted the need for a tripartite approach involving the community, the regulator and industry in the development of government policies, guidelines and procedures. The Wagerup Tripartite Group (WTG) was subsequently established in January 2004 by then Environment Minister Dr Judy Edwards to address emission licenses and other relevant environmental issues surrounding the Wagerup facility. The WTG forum promised to be open and transparent in its communication with the local community. It was to be run at arm's length from government and industry through independent process facilitation [55].

CCN and WTG meetings continue to be held at regular intervals, and the minutes from these meetings are made public [see for example 56; 57]. However, despite the apparent commitment by industry and government to due process in terms of community consultation, neither CCN nor WTG had the approval of key community stakeholders due to perceptions of bias, lack of autonomy and insufficient decision-making powers. Thus the principal process devised for dispute resolution was largely boycotted by community members and as such failed to capture views that were critical of government and industry processes.

APPLICATION TO EXPAND

Amid ongoing conflict, Alcoa applied to expand the

Wagerup refinery, proposing to almost double its output capacity [58]. This fuelled community concerns about worsening emissions and community impacts [59] and triggered considerable opposition [60; 61]. Independent members of the Wagerup Medical Practitioners Forum also argued against the proposed expansion on the grounds of unresolved health issues [62].

The evidence arising from earlier studies clearly indicates that the geography and topography of the area was never suitable for the placement of an alumina refinery.

The evidence indicates that some of the neighbouring community members in our professional opinion have suffered acute and chronic adverse health consequences as the result of their close proximity to the existing refinery.

... there has been insufficient duration or consistency of an improved performance on this issue to give us confidence that Alcoa has accepted ownership of the problems.

Where responses have occurred to the health problems, these have generally been unsatisfactory.

There are considerable uncertainties surrounding this public health problem ... No-one can model with sufficient certainty what the short- and longer-term health consequences of expansion would be. The most relevant point of certainty is that the history to date has been one of adverse health effects for community members.

This led the forum to the following conclusion in 2005:
In summary, we do not support the proposal to expand the Wagerup refinery in the existing circumstance of an inadequate buffer zone. Our judgement is that, in the face of much uncertainty, the problematic history

> *of the relationship between the refinery and the local community is the most reliable guide to what the future would hold if the refinery was to expand. On this basis we consider that the risk of further compromising the health and social functioning of the local community to be too high,: and the trade off of this risk against the broader benefits to be unjust.*

Notwithstanding, following the approval from the EPA [63], the state government granted Alcoa's expansion plan in 2006 [64]. With the recent global economic downturn, however, the company has put its expansion plans on hold [65].

ADDITIONAL BUY-OUT: A (LATE) GOVERNMENT INTERVENTION

In 2006, the state government announced the establishment of the Supplementary Property Purchase Program (SPPP). This program was an agreement between the government and Alcoa, and formed part of the government's environmental approval for the Wagerup refinery expansion. The company agreed to fund the program which would enable eligible property owners outside land management areas A and B to sell their properties to Alcoa. Hendy Cowan, former Member of Parliament, was appointed as the administrator of the SPPP.

Officially, the company did not expect strong public interest in the SPPP.

> *I don't think [the uptake] will be huge. There are a lot of properties in there. I guess my interactions with the locals over the last few years have been that the vast majority of people are pretty happy. The majority of people are pretty comfortable and happy to stay there and want the town to go ahead. (Alcoa management staff)*

And nor did the government [66].

The interesting thing about the SPPP is that ... the government was expecting a handful – probably 30 property owners – to put up their hands and say that they would like to be moved out of the area ...

In fact, the government was inundated with requests by locals to have their properties registered and valued under the SPPP. By mid-June 2007 registrations for 408 residential and farm properties had been received, causing delays in the processing [66].

The government scheme was criticised for its slow implementation and for other reasons. Residents were given only seven months to register with the SPPP, three months to accept an offer, and four months to settle. Established to cover communities outside Alcoa's land management areas, it excluded properties inside areas A and B. Residents further afield wondered why they were now considered to be living inside a special land management area designated by government. Furthermore, the scheme had no provision for the waiving of stamp duty, nor was relocation money paid for residents who chose compensation, creating large out-of-pocket expenses for community members. Finally, residents felt that their properties were undervalued, being rated lower than comparable properties in nearby towns because of the refinery stigma. Many concluded that the SPPP was even less generous than Alcoa's purchase program in areas A and B.

Attempts to initiate a government inquiry into the fairness of the SPPP were quashed in the WA Senate [66] under what can only be described as bizarre circumstances. A media release by Paul Llewellyn MLC (the Greens) captures the event.

Parliamentarians turn their backs on Yarloop residents
Parliament witnessed one of the most hypocritical, stage-managed backdowns in its history, when the Liberal Party, with the full support of the Nationals and the Labor government, voted against their own motion which called for an inquiry into the fairness of the Alcoa property purchase program.

Residents suffering from the harmful impacts of the air pollution around Alcoa's Wagerup alumina refinery have been pleading for a fair and just exit deal to move away from the area. The Supplementary Property Purchase Program (SPPP), announced by the Premier when granting Alcoa approval to expand its operation in 2007, has been plagued by delays and accusation of unfair dealings. In a face-saving exercise last year, Liberal Upper House Member, Robyn McSweeney, put forward a motion to conduct an inquiry into the fairness and just terms of the SPPP. Today she and her colleagues voted against her own motion, effectively abandoning her south-west constituents and putting her support behind Alcoa. This cynical manoeuvre shuts down community access to independent assessment of their concerns by the Parliament. [67]

COMMUNITY MOBILISATION AND A CLASS ACTION

In the dozen years from the mid-1990s the town of Yarloop experienced dramatic changes. What local statistics and anecdotes portrayed as a thriving town [68; 69] changed into a dying place [70]. The Yarloop community lost most of its local businesses, including two petrol stations, the local shop and the town hospital [71]. Population levels dropped by 45 per cent and property values fell sharply [69; 72; 73; 74; 75]. Despite numerous studies, the Parliamentary

Inquiry, extensive media coverage and select government interventions, many of the issues persist. In the eyes of one local resident:

The current situation is exactly the same as it was in 1996, given that the impacts are ongoing, and whether it's noise, dust, odour or health – and it is still all four – it hasn't changed! It has actually got worse ...

Problems went beyond the questions of health and finances which featured prominently in the media. Much of the conflict centred on personal and social aspects, the very core of family life and community cohesion and wellbeing. For residents it was not primarily about compensation, money or property. Instead, it was about what Yarloop and its community meant to them, the loss of which could not easily be compensated.

Over time various community groups formed, voicing concerns and opposition to what they experienced as severe and far-reaching changes to their lives and their communities. The Yarloop and Districts Concerned Residents Group and the Community Alliance for Positive Solutions Inc (CAPS) have campaigned for many years to hold the state government and Alcoa accountable for the negative impacts inflicted on individuals and community life by the refinery.

These groups were active in making public submissions to government and presentations to ministers and government departments. In recent years, they have organised community workshops to facilitate information sharing and community empowerment. The public profile of the Wagerup controversy owes much to the work of community members who, by way of newspaper advertisements and media reports, successfully raised awareness about the experiences of residents with the refinery. Publicity was a vital aspect of the campaign in light of Yarloop's relative remoteness from the metropolitan area and media spotlight.

Today, CAPS is the most vocal and active group in Yarloop with over 400 members. It has developed into an internationally networked organisation in pursuit of justice for their community (its slogan is 'We were here first') [76].

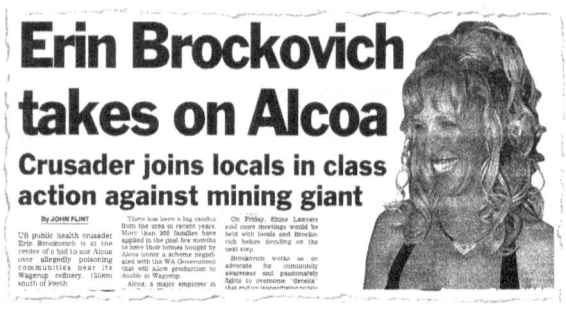

Media coverage on class action against Alcoa [77]

In 2008, with the help of environmental campaigner Erin Brockovich, a multi-million dollar, high-profile class action was announced against Alcoa by residents [78]. Legal proceedings began in 2009 in the US court system. The writ accuses Alcoa of 'knowingly, negligently and recklessly operating its factory and poisoning surrounding communities with toxic emissions' [79].

Alcoa also faces court in WA, charged by the Department of Environment and Conservation for an alleged pollution offence at the Wagerup refinery. Initially the charges were alleged pollution with criminal negligence, but these were downgraded in 2009 [80; 81]. At the time of writing both cases are ongoing.

CHAPTER 3
LOCAL STORIES ABOUT REGIONAL (UN)SUSTAINABILITY

In this chapter the residents of Yarloop and surrounding areas tell their stories about life under corporate skies and share their understanding of regional sustainability in terms of the issues most important to them: people and place. Their narratives weave together the many different experiences of the changes felt by individuals living near Alcoa's Wagerup refinery between 1996 and 2007. Together they show how a company's activities and processes have threatened local health, safety, stability and undermining the very sustainability of their community.

We have given voice to as many people as possible, however space constraints prevent us from giving stories in their totality. We therefore organised them under themes that capture the essence of the Wagerup controversy and give structure to the many and varied accounts.

An important aspect of the conflict is the longstanding ties many residents have – or had – to the area. A significant proportion of Hamel, Cookernup and Yarloop residents lived in the area prior to the arrival of Alcoa. Some families have lived in Yarloop and neighbouring towns for many generations, and have strong emotional ties to the place and its people.

My mother was born there, my family was born there and my father came down in 1912, and Marie's father came here when he was six months of age, so we are very old identities, our families are. My uncles came to Yarloop in 1888; so we've got a whole connection.

Dad's family came there in 1910–1911. Mum was born in Yarloop. Her family came in 1906 so more than 100 years of history we have associated with that town.

Yeah, my grandfather was there and my great-great-grandfather and great-grandfather. Yeah, it goes back a long way.

A number of residents came to the area when Alcoa commissioned its Wagerup operation. Before the conflict

I came here as a child (Photo by H. Seiver)

began there had been a steady growth in 'lifestyle' settlers who were attracted to the area for its sense of community, its tranquillity and its idyllic setting between the Indian Ocean and the Darling Scarp.

> *... We always used to come down there ... to spend our weekends ... because we had a pretty stressful sort of a job situation in Perth ... We liked to come to this idyllic place. When our job situation changed – the company decided to shut down and we were made redundant ... we decided to take the 'treechange', [get] rid of the mortgage and come down here.*

> *It had everything I needed. School, hospital, space, shopping, not far ... from the train if the car was unreliable, because my car was ageing. I actually dropped out of the rat race, and it was close enough to Mandurah and Bunbury for going to the theatre and shops or specialist doctors, and not too far from Perth if you wanted to go for some special thing in the city.*

> *I like the lifestyle and it's peaceful. I just really fell in love with the place and I really don't want to live anywhere else.*

In fact, Yarloop was described as a much sought-after place many people were attracted to:

> *It was a real lifestyle place and in order for someone to come to town, [you] virtually had to wait until someone died to get a house or a block here.*

With this influx of newcomers, Yarloop experienced a steady growth in population and residential developments, comparable to other small towns along the inland highway. In the eyes of its residents, it was a place with 'a very close-knit community', strong social cohesion, long family histories and a strong sense of place.

Children of Yarloop (Photo by H. Seiver)

It was an absolutely wonderful little town. It was a great place to live in, no problems whatsoever.

The community at that point ... was very close-knit. People would help one another; if someone was in need or needed financial help or had problems, someone would come along and give you a hand.

If someone was in strife in Yarloop everybody was in strife in Yarloop. If you had a problem at your place, before you knew it there'd be a dozen blokes here, a couple of bottles of beer and they'd get your problem fixed, have a few drinks and all go home again and think nothing of it.

You knew everybody. As the kids grew up you didn't really worry about where they were because you knew everybody; everyone was looking out for them. (former resident)

Yarloop today is described by the same residents as being 'totally different now, because ... a good 80 per cent ... of the original residents have sold up and left the town.'

(Photo by H. Seiver)

These days, life in Yarloop is considered 'scary' and 'horrible'. The place is seen as a 'ghost town' where 'the quality of life ... has just disintegrated.'

Dead! It just doesn't seem to have any life any more.

A shithole! Honestly, I could not live there ... the atmosphere, what you can see in Yarloop, it has deteriorated that much. (former resident)

... Well it's just almost no town now. I feel it's just a dying town.

Over the space of only ten years, according to its residents, what was once regarded a 'vibrant little town' was now likened to 'Wittenoom Gorge'.

It's a beautiful place ... but I wish to God I had never moved here. (former resident)

It's gone from the penthouse to the shithouse, there's no doubt about it. (former resident)

HEALTH IMPACTS

Health was a key aspect of the Wagerup controversy, relating to the emissions from Aloca's refinery and their perceived impact on community members. Health concerns started to emerge in late 1996 following the commissioning of the liquor burner.

To me it's all linked to the liquor burner ... Once that liquor burner went in that sort of, I don't know, just made things worse. People got crook. It seemed to be once that went in more people seemed to be getting crook.

While health impacts are well documented and are generally available on public record [1; 2; 3], an overview is presented here to convey community concerns about the refinery and the circumstances of individuals living close to the plant. Initially, residents became concerned because of unusual experiences they had when outdoors, especially during winter months.

In the morning I go across there and walk my dog down the railroad track and back, and if the wind's from that direction, you'd think you'd just come from your best friend's funeral. Eyes just water ...

... It started to rain and ... my hair was falling out and I was getting really bad stomach pains ... [M]y feet

were just burning. I would get these sore throats and ... go downhill real fast like chronic fatigue.

Every morning I went out to get the cows in the middle of winter I used to cough until I vomited. It wasn't a good look, and you know my son and I ... we could just be standing there talking and all of a sudden he'd get a blood nose. (local farmer)

For many residents, health problems were noticeable in the form of flu-like symptoms, affecting the throat and breathing. Other symptoms included:

... Headaches, burning in the eyes ... on the skin, being nauseous [and] really tired [and] hair loss.

For some residents the symptoms were relatively mild and transient, but others experienced severe long-term problems after exposure to what they believed to be Alcoa's refinery emissions.

... My skin, I get burnt. It's like a radiation thing ... You also have bladder problems and it affects your bowel, it affects your moods, it affects your skin, see my skin is horrible. I can't explain; my stomach is always sore after I've been outside and stuff has come on me. Sometimes I am just irritated like I want to scream and I am saying, 'Oh what is wrong with me?'

Then it got really bad in 2000 when ... I thought I was going to die that day. It was January, stinking hot, and I had a doona cover to keep myself warm, I was freezing. I was in bed for five weeks and I thought I was going to die. Just lying in bed, I couldn't get out of bed for five weeks ... [I] had a lot of tests done and then [I] went to a naturopath and he told [me I] had chemical liver poisoning, that [my] liver was not coping with all the toxins in [my] blood.

My organs were shutting down they just couldn't handle the toxins any more. (former resident)

A number of residents are suffering to this day from illnesses such as chronic fatigue and multiple chemical sensitivity (MCS). Medical conditions such as these are difficult and controversial in their diagnosis, and sufferers battle not only with their health but with the legal and medical professions to have their condition recognised.

Many of the residents' encounters with what were described as toxic plumes occurred outdoors, making

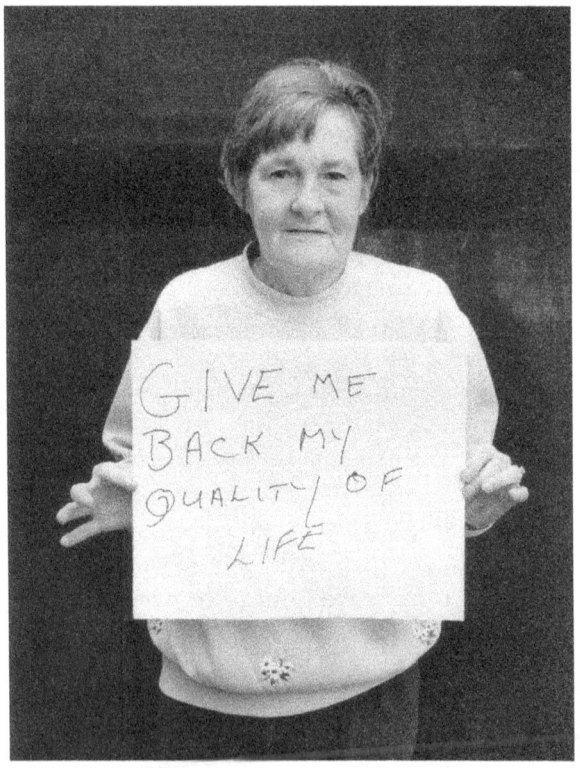

(Photo by H. Seiver)

locals more cautious about spending time outside. A sense of entrapment developed as residents felt they could not leave the house. Some became affected by the emissions in their own homes, leaving them with nowhere to go.

> *A bit of a breeze came through the bedroom window, and this breeze come through and it stunk. It smelt like pollution. It burned and I jumped up ... ran out on the front verandah, stood out there and I just got burnt by the plume. I saw it. It was coming down the railway line and it was like a greeny, dirty, greeny brown. I just went outside and I just got burnt from head to toe ... my eyes, mouth, nose, the whole bloody lot. I thought, this is it, I'm in a nightmare. I couldn't believe it. (former resident)*

> *[At] night times you'd be sitting in your lounge room and your eyes would start to burn because it used to come through the bloody floorboards, so you'd have to lock all your windows up. It was like ... you were literally a prisoner .... (former resident)*

Some residents showed no signs of ill-health, but were anxious about the changes in health they observed in others.

> *I have friends with grandchildren who were conceived locally who have major medical problems which were not systemic or endemic to the family; they were something new. I've got friends who are coughing and choking and saying: 'there's nothing wrong with me ... cough, cough, cough.' You know ... 'Good morning, cough, cough!' That's Yarloop these days.*

Problems were also evident in their pets and livestock.

> *Our cows and calves would end up with eye infections and all sorts of problems. (local farmer)*

> *My cat, when we came here, she started getting this bloody dermatitis, this infection.*
>
> *The dogs are always getting tumours cut off them.*
>
> *The woman who first stirred it all up ... reckoned they could not get the cows and heifers in calf. [The neighbour] said, 'Oh that's rubbish,' but later on one of his sons said, 'I'm having a bit of a struggle getting a good acceptance rate of our heifers', so his had dropped right off, too. (local farmer)*

Initially, health symptoms were considered isolated, as people were not all affected equally or simultaneously. It was quite possible for members of one household to display symptoms while next-door neighbours did not experience any. These uneven health impacts gave rise to friction between people with health problems and those without. Nonetheless, over time, many residents started to realise that symptoms were shared widely throughout the town, and suspicion grew about links between community health and the Wagerup refinery.

> *Everybody I talked to had the same problems as me ... So ... in a street of nine houses, seven people were affected. It was just horrible, and you'd go across to the shop and there's [the owner] over there, he's crook, you know.*
>
> *I attribute it purely to the emissions from Alcoa causing problems. I have lived in the district now 40 years ... I'd never heard of any health complaints. All my friends that lived there were happy, healthy, never made a complaint about either their health or the company or the town or anything, so I attribute the health effects purely to Alcoa.*

Unsurprisingly, health featured prominently in residents' complaints about Alcoa. However, the issue proved to have more far-reaching impacts, affecting people's longstanding ties to the area, their families and friendships and leading over time to a domino effect which served to dismantle this very close-knit community. Health became a trigger for a series of impacts on the community felt also by those who did not suffer any ill-effects from the refinery emissions, as poignantly summarised by one local resident:

In one sense or the other nearly everyone in Yarloop was affected, maybe not by the issue of the plume or the smell but by their friends leaving or they don't agree with what's happening with Alcoa and they don't believe that it's Alcoa but their kids are being affected so it's causing a rift between the family.

QUALITY OF LIFE

Acute health problems or the fear of falling ill meant that the daily lives and routines of residents became affected by the presence of the Wagerup refinery. Residents spent more time inside, out of fear of exposure.

Who wants to live their life being in a house where you can't walk outside because as soon as you do your eyes are going to water or your throat is going to burn or your nose is going to bleed? People shouldn't have to live like that; it's a free world and you should not have to live like that. (former resident)

I didn't come here looking for that, I came here to live in a house and I live here; actually, I exist here, I don't live. I don't enjoy my house and that's tragic because it's a great house but I can't enjoy it.

This meant they also refrained from pursuing outside activities and hobbies they used to enjoy.

... All my activities outside – mowing lawns, gardening

– I love it. I absolutely love lawns and gardens. That all stopped and my husband had to do it all himself. Orange trees, we had heaps of orange trees there; big orchards. No more picking oranges or anything else, I just couldn't do it because I would suffer. If I did, I would suffer. (former resident)

... When I go walking I break out in this rash, every time. I don't go outside hardly, I've put on weight. I didn't exercise, I didn't do gardening, mow lawns nothing, I stayed inside for probably about two years out of my life. I had not gone outside very much at all. All my outdoor activity with gardening and that all

(Photo by H. Seiver)

> *stopped. (former Wagerup resident)*
>
> *So that lifestyle was interrupted because once we realised or found out what we were actually being affected by and what the ramifications of those chemicals were, we virtually stopped growing vegetables. Our chooks kept dying so we thought if they're dying it's not really good to be eating their eggs so we stopped that. Bit by bit we just stopped doing everything that we were doing for our lifestyle by choice because we just decided it wasn't for our best interest. (former resident)*

There were other social costs, as people stopped seeing friends and neighbours out of fear of being affected, or residents would worry during social engagements that they would be embarrassed by the refinery impacts.

> *People would come down from Perth to visit and straight away they'd say, 'Oh that smell, how can you stay, how can you smell it?' (former resident)*

Olfactory stress was compounded by noise from the refinery as well as the train traffic. The rail line connecting Perth and Bunbury runs along the South West Highway and right through the centre of Yarloop. Alcoa uses the line to transport caustic and alumina to and from the refinery. With rising levels of production came a rise in train traffic.

> *On a busy time on a line here we can have up to seven or eight trains an hour coming and going …. It's fairly busy and if every one blows their horn and all that sort of stuff, it gets annoying after a while because if you're outside, you've actually got to stop your conversation.*

Both the noise and the smell of the refinery were cited as having big impacts on the lives of residents who could not protect themselves from exposure or from unwelcome trade-offs of protection.

> *We could not sleep at night so they came and put double-glazed windows to stop the noise so in that way we couldn't open the windows at night on a hot night because of the noise. We would be up two or three times in a night ringing up. If it wasn't for the smell during the day it was the noise at night. (former resident)*

> *We used to sleep with the window open in the bedroom and now the last few years we just have to shut it because of the noise and the smell in the air. You have to just shut yourself in the bedroom. (former resident)*

> *The smell was so putrid at times my clothes on the line would stench. So I used the dryer all the time ... (former resident)*

The refinery became a central feature in the lives of residents. They started to feel they were living 'in the middle of an industrial zone'. The far-reaching impacts on people's lives were described as 'devastating', and residents expressed their resentment towards the company they felt was responsible for their loss of quality and enjoyment of life.

> *... It hits at the heart of your lifestyle. It's a dislocation. I think it's a total dislocation into our lives, the people that are affected. Hurt because you've been affected by health and hurt, the fact that our lifestyle and tranquillity of where we are and been here for generations ha[ve] been in upheaval .... (former resident)*

> *I have lost a great deal of my enjoyment of life.*

> *Yep, everyone's angry with Alcoa. Everyone is really angry with Alcoa and they're angry because they've taken away their lifestyle.*

I'll hate them forever. (former resident)

THE IMPACT ON FAMILY AND FRIENDS

Most community members cited the toll the Wagerup conflict took on families. Health problems or the fear of being affected by pollution led to much family friction, the uprooting or the separation of once close family units. Many of the families in town, especially many of the Italian migrant families who came to Australia following World War II, were living close to each other or even next door to each other on subdivided large properties:

We had 13 to 14 acres of land ... The blocks were subdivided for the kids to build ... That's all gone down the chute. (former resident)

Mum was in an ideal situation in the sense that she had, of her six boys, four of them ... liv[ing] within walking distance: two on either side of her and a couple of others scattered around. For her, even though the family married and had their own families, that connection was still very strong but that was pre-1996. (former resident)

For health reasons, many family members, particularly the elderly and those with young children, left town, resulting in the separation and disjointing of families.

The families have all broken up. It was a very 'family' community and everyone knew each other and that's not there any more, it's all broken off. People have split up and gone to different towns, different places due to the threat of pollution from Alcoa and future expansion from it and people have moved on. (former resident)

Families that once enjoyed closeness and regular contact found themselves living in different towns with diminished opportunities to see other family members.

A lot of the people have shifted out. Of all the people [my wife] grew up with, all her uncles and aunts, and friends, most of them have shifted out now. Gone in all directions, down to Harvey, Waroona, Mandurah.

... There is a photo up there on the wall: that is five generations of my mother-in-law's mother and I would say 90 per cent of those people lived between Waroona and Yarloop up until a few years ago. Probably 95 per cent of them are family who we have now had to move away from. Ninety per cent of them were all between Waroona and Yarloop. When you look at that it really hits you. All those brothers, sisters, brothers-in-law who saw each other every day ... walking and seeing each other every day, and now they're in all different towns. We hardly see one another now.

My husband is still angry. He's so angry – very, very angry – because being an Italian the tradition is you give the kids a block and they were going to build there and now the kids are scattered; one's in Australind, the other one's in Galway Green and my daughter ... she went to Perth. So it has split up a whole family, a close-knit family. I'm not just saying a close-knit town; this is just what it did to our family and many, many others. (former resident)

Residents who remained in town were concerned that their own families would also be affected, having witnessed the break-up of many family units around them.

One thing I've got a fear of is that it might break our family up if we have to move. If we have to move it could split our family up; one will go one way and one will go the other way; we'll lose that family thing that we've had all our lives. I wouldn't like that to happen. Like it's happened to other families in this town that have had to move; it's pretty sad.

Others lamented the fact that due to acute health issues or health concerns their contact with family members was no longer frequent, as people were becoming less inclined, even afraid, to visit Yarloop out of concern about what it would mean to their health:

My grandchildren no longer come here because they get sore eyes, itchy skin, blood noses and all the classic symptoms of the effects of Alcoa, so my grandchildren won't come here.

My daughter from Rockingham hates coming here. She said the kids have got to play out in that yard.

Because health impacts were felt unevenly, even within families, arguments arose as to whether or not it was safe for the family to remain in the Yarloop area.

It caused a lot of problems for me and my partner. It caused massive arguments. (former resident)

... It's causing a rift between the family. It has caused divorces, more than one or two. It has split family against family.

I'd get to the point where I feel very lethargic ... and I'm having this symptom and that symptom. My kids were quite small at that stage ... and my wife was saying, 'Look, we need to get out of here.' I thought: 'I've been here all my life, why should I get out because of what Alcoa is doing?' So that started the ball rolling to the marriage ending.

Well it was either [move] or get a divorce ... In the end, [my partner] said we either sell up or I'm out of here, and I couldn't blame her. She'd had enough. (former resident)

The effects on families were also felt further afield with the exodus of many long-term residents between the late 1990s and mid-2000s. Not only were family relationships affected but also friendships across town.

So my loss is the friendship, the connection I've had with all the friends and people I grew up with. Now there's probably only a handful left ....

All our friends now have moved out of here and we don't know anybody that comes down that street. When you meet somebody in the street you don't even know [them]. Years ago we knew nearly everybody in the place.

The social connection, the friendship, the people looking after each other, the way this town was close and worked together. That part of it there is gone and that's what I miss the most ... Yeah, it isn't the same town I used to know.

SEVERING LOCAL TIES

There was a widespread sense of emotional loss about the people's local history. Many residents were grieving the loss of longstanding family ties to the area. As residents started to move away, local family histories became fractured and family traditions discontinued.

I was third-generation Yarloopian and [my children were the] fourth generation ... in Yarloop. There will never ever be a fifth generation ... in Yarloop. That's my history gone. (former resident)

... there is an emotional thing in moving from somewhere where you've been – you've had generations of family that have established what you've got; that's

hard to replace, it can never be replaced. (former resident)

The adjustment and change for us has been huge as the family had lived in Yarloop for four generations. (former resident)

For those older residents who had been living in Yarloop all their lives, having to leave town for health reasons was particularly difficult and painful.

... my mother-in-law, she's 93, she was crying, she could hardly walk ... and she was kissing the house goodbye. My sister-in-law was kissing the house goodbye, everyone was crying. That was the most pitiful thing at 93 and 94 to leave their little house and move ... they've never got over it and the anger they still feel, especially my mother-in-law. She will never forgive Alcoa, never in a million years. (former resident)

That was our life. That was our whole life, our kids' life, our life. Now it's just a sad story and you know what, it will go down in history. (former resident)

Most of the people who moved away from Yarloop left reluctantly. Many had lived there long before Alcoa arrived, and were expecting the company to clean up its emissions.

Now, we were here first. We're not polluting Alcoa, but Alcoa's polluting us. (former resident)

They destroyed, they intruded on my life; I didn't intrude in theirs. (former resident)

In the end, these residents felt they were forced into relocating due to poor health. For those concerned, it was not a free choice since staying true to principle became, in their view, a matter of risking their lives.

> *Well, we were pushed ... we were basically pushed out. That's how I feel. We made that decision to go and get away from it ... but we were forced into that decision. (former resident)*

> *Well, I suppose you do feel pushed out. Pushed ... Well, I guess we had the option: either you stayed there and be crook, or you moved. So I suppose you are pushed. Otherwise we would still be there. (former resident)*

> *We liked the area ... Our kids love it; they don't really want to leave; we don't want to leave; we feel like our hand is being forced.*

> *My son and his wife, they had two little kids and the oldest girl is eight, she was chronically sick ..., stomach aches ... it's the health issues that forced us to actually go, not because we wanted to, we were forced to go and that's the only word I can put it down to is 'force'. We didn't want to go but Alcoa wasn't going to go so we had to go. (former resident)*

> *There are no willing sellers here; they're selling because they have to.*

Leaving came at an enormous emotional cost as those who left felt guilty about having 'let down' the community. They also felt defeated in having lost their 'battle' against the multinational.

> *I felt I had no choice. It was either that or go mad but I know I hurt people when I left. I let them down ... (former resident)*

> *We have done everything in our power, everything to try and do the right thing and all that we had to do. Alcoa won: we moved out. (former resident)*

I know people here that have been fighting for ten years and you have to come to a point where you've got to wake up to your own life and do what's right for you and I think standing up for your rights is the right thing to do but to what level of expense personally do you go? Is it your health, your wallet, your future? (former resident)

LOSS OF COMMUNITY

A strong sense of community and social cohesion were key characteristics of life in Yarloop. Longstanding ties with people and place meant that lasting relationships could develop over many years, resulting in a community where 'everybody knew everyone in town' and 'where people helped one another'. Community dynamics changed with the onset of the Wagerup conflict, in part because people started moving away.

It started to change once people started to become aware of what was getting pumped in the air from the liquor burner. I saw a change from that. People were starting to get worried and one by one you start to see people move out of town.

Among those remaining, there was disunity over the cause of the emerging health problems. Some of those whose health was unaffected could not understand what the fuss was about.

I really don't know. I really don't, because it's never really been proved that the emissions are that bad. I mean, all right, you've got your handful of people that would be allergic to what comes out of smoke but you get that with peanuts and other things. There's always a few that get really affected and I feel sorry for them and they deserve the help to move if they can't put up with it.

Sceptics denied the existence of problems as health effects seemed isolated and rare. In their view most people appeared to be fine.

Most people you talk to have no problems; either people we know in Hamel, Waroona ... They've got no problems at all.

In turn, sceptics were accused of being in denial.

The common saying ... was: 'it doesn't affect me and I don't want to know.' (former resident)

That has been their position: 'It doesn't affect me and I don't want to know' and they hope that whilst ever they keep denying it, it won't affect them ... (former resident)

Some of those affected felt hurt that their health claims were not believed.

[Some people say] ... yeah, you hypochondriacs, get out, hurry up and leave so we can have real people in here and that is so hurtful ...

As the conflict intensified they felt they were being stigmatised for being sick.

When you find out you are part ... the 'affected community' ... you are being shunned. We've had some social shunning going on pretty heavily, when you start to say you are sick. (Cookernup resident)

DIVIDE AND RULE

Community friction increased following the release of Alcoa's Land Management Plan which differentiated between residents in areas A and B and those outside the management plan areas.

[Alcoa] split the town. It's like they've drawn a line down the middle and they've said, 'You've got brown eyes, you're on that side; you've got blue eyes, you're on that side.' ... They've made everybody argue and fight. Everybody bitches and fights about how come they get more money than them and why aren't they getting it. Everyone has got an argument. So it's not a friendly town anymore. No one's happy.

The Land Management Plan changed the nature of the growing friction in Yarloop, which became increasingly focused on money and property values. This injected a venom that had not been seen before and deepened divisions in town. The photo below reflects the intensity of the community conflict.

Community friction: a low point (Photo by unknown local resident)

Those claiming to suffer from refinery-related health problems and who blamed Alcoa for destroying their community were accused of being 'troublemakers' and of seeking higher pay-outs and compensation.

... you don't even complain anymore because people look at you and they pretend they're listening but deep down they probably think 'What a lot of crap, you're

> *making it up just for money or some benefit', so you don't really discuss it with anyone.*
>
> *Look, the only problem in Yarloop ... there's a bunch of troublemakers in Yarloop and they are trying to make things really tough for Alcoa. But there's nothing wrong here.*

Those who spoke in defence of the company said they enjoyed a good relationship with their corporate neighbour, who they trusted to have refinery emissions and associated pollution problems under control.

> *We get on very well with them; they are exceedingly good to us.*
>
> *There are other people who are worried. I don't believe they've got any reason to worry because we've been made a promise that there'll be no more emissions, no more noise, no more whatever else and they've got to adhere to that to be able to keep their license, so I don't think we'll have a worry.*

Those critical of Alcoa, however, believed that pro-Alcoa residents were either ignorant of what was occurring in Yarloop or benefiting financially from 'sticking up' for the company.

> *So there's that sort, where I don't think there is any gain there at all from Alcoa other than they are getting poisoned and they don't know it. They don't care. Then you've got the other ones that know what's going on and that are there because they're making money out of Alcoa and they are the ones that do a lot of damage.*

Alcoa supporters believed that the company was unfairly targeted by a noisy minority that seemed to blame every personal and community problem on the refinery.

> *Everybody seemed to be blaming and they're talking about emissions and noise and I couldn't hear it, I'd never smelt anything ... So there were all these things that you couldn't believe most people that were shouting the loudest had these ulterior motives, and you just couldn't believe most of it that was going on.*
>
> *I don't hear it from here. I don't hear any noises. The only noises I get is from the wood factory at the top there, that I can hear. So whether people are confused that it wasn't Alcoa and it was the wood factory, but you see, where do we stand? We had Cable Sands and we had the wood place and Alcoa, I mean, you've got three, you can't just blame one, you've got three and the industry is important and people should understand that.*
>
> *See, closing the police station down and closing the hospital, everybody is pointing that it's Alcoa's fault. I mean, is it or is it just that there is not enough funding to run such a small town?*

Those in support of Alcoa did not share the view that Yarloop's future prospects were bleak. Indeed, many expressed a strong sense of optimism.

> *This town will grow big-time. There's just been a subdivision up School Road, up behind the houses there ... The town is growing.*
>
> *[Yarloop's] got so much potential and it will move forward, regardless of the whingers at this time, this town will move forward. It will probably be tourism that will keep this town going.*
>
> *We need to get on and get our town going again. [Look at] the workshops; they're just going ahead in leaps*

> *and bounds and now we have the town beautification scheme coming on ... that's going to do wonders for the place. The town is moving ahead regardless of this mere handful of people who would like to pull it back.*

The optimists believed the 'whingers' and 'complainers' were the problem and that it was as a result of their concerns that Yarloop was thought to be going backwards and property values were declining.

> *They're sticking back in the dim dark ages. They're listening to old fairytales and they don't want to move with the times and they don't want to know and you can't do anything to help people like that.*

> *If you don't sell up, shut up; because you've got your opportunity to get out. Before you reckoned you couldn't sell and half of it, as far as I'm concerned, is their own fault because they've put nothing but bad things in the press and then they've said, 'But I can't sell my house.' Well what would you expect when you're giving bad press all the time!*

CONFLICT FATIGUE

Another group of residents simply wanted to move on and let go of the conflict. While they did not experience any problems with the refinery per se, they felt affected by the derailment of community dynamics and the negative publicity. They were tired of the animosity and just wanted to get on with their lives.

> *I loved it here and I think I still would if they would leave us alone. I think it's time to stop all the publicity. We should be left alone.*

> *The newspapers. We are always getting them. I mean they closed the hospital down and we get front-page*

news and it's just, it's time to say good things about the place and stop.

The Wagerup conflict pervaded all aspects of community life and eventually nobody could escape from it. Over the years, a fatigue set in across the community.

It dominated everything. Exactly, that's what it was. We couldn't go to a party. It was horrible. You couldn't go out without the conversation of Alcoa, complaints, this and that. (former resident)

You can't have a conversation in Yarloop without Alcoa. I can guarantee that. It's overwhelming. It's what you eat, you breathe, you sleep, you dream. It's how you look at life. It's how you perceive everything. You become so angry.

This sense of fatigue was also discernible among those who fought against Alcoa to protect their community, properties and lifestyles. After years of writing letters of complaint and attending countless community meetings, they found themselves worn down and giving up. In the eyes of some, community burnout was employed as a deliberate corporate tactic – the company could rely on the fact that community members would give up eventually.

I'll probably just get up one day, say 'stuff it' and piss off. I'll take what I can and go … because I'm really sick of this whole thing.

Oh, God, what can I say? I'll be glad when it's over. I've got to that stage where there's only so much fighting you can do

… they have a game plan to wear people down, so that in the end you will just throw your hands up in the air and say, 'I've had enough, okay, whatever, I'll sign it.' (former resident)

> *That's what Alcoa wants, that is what they are wanting: for us to give up and to stop complaining.*

But as ongoing court proceedings attest, the conflict continues.

> *I'm a fighter. I am not gonna give in ... I think they will really fight this case with us, and I will keep fighting.*

DEMOGRAPHIC CHANGES

Friction was also a product of 'demographic change'. When long-term residents started to leave town a new cohort of residents arrived, presumed to be attracted by low rents and property prices.

> *So they've brought in an element ... you've got people with low social economic ... they've got no money. The whole area has turned into a bit of a quagmire. If you knew who was living around there, you wouldn't want to live there. (former resident)*

Remaining long-term residents felt that Yarloop was 'no longer [the] family environment' it used to be. It was described as 'a transition town now' with 'no locals ... they're all strangers'. In short, 'it [wasn't] the same town [they] used to know.'

> *You just don't know who's around ... Now, it's just sad, you know, you don't know anyone and there's just nothing here anymore, there's just nothing. (former resident)*

> *At one time everybody here had been here for years. Now you get these transient people who come in and you don't actually get to speak to them or say hello because they're gone the next day.*

With the influx of newcomers the nature and appearance of the town changed. The place started to look and feel different, clashing with the traditional values held by locals who prided themselves on their well-maintained houses and yards and their involvement in community life.

> *Well you just drive up the street there and have a look at the houses there that are derelict. With old bodies of cars sticking out the front there it looks like Steptoe and Son.*

> *There's no heart and soul left here anymore, whilst there are a lot of people still here – and there are probably just as many people as ... there was before, a lot of renters, people have bought into the place ... (former resident)*

> *People have got no respect. There's a mentality there that's living there at the moment and it's scary.*

Locals also felt that 'drugs and undesirables ... entered the township' and that 'crime increased'.

> *We used to go away and leave the house open. You don't do it now! Got to shut everything.*

> *They bought up some of the properties and without screening people they put them in. And we know there are drugs at the top of the street. Several times we've had the police up, arresting people, crimes. So that's taken the sort of easy lifestyle away from the people who live here, because you don't know who's in the street or right next door to you.*

Newcomers were treated with suspicion by long-term residents, who saw their town and its values changed and deteriorated. These negative reactions to change can be seen as a characteristic of the so-called 'dark side of social capital' found in communities where social cohesion is accompanied by hostility or prejudice towards outsiders, and strong expectations on social conformity [4; 5; 6].

However, beyond the parochial sense of 'stranger danger', newcomers were also seen to dilute the cause of long-term residents because of their different relationship to the town and their apparent reluctance to get involved with community activities and issues.

> *The heritage is all lost now because there is no one there that understands Yarloop. (former resident)*
>
> *... new people don't volunteer for anything. Yeah, look, they're very reluctant to talk and even just say hello, you know what I mean? They sort of ignore you.*
>
> *The rentals, they're just not interested.*

As newcomers started to outnumber long-term residents, community opposition to issues such as Alcoa's proposed expansion or the company's land management plan lost strength in numbers.

> *It was difficult to maintain community spirit in that sense, like let's stick together and fight this together ... I suppose the community strength had started to deteriorate ...*

We also found in our research within the community that new arrivals were not keen to comment on the Wagerup conflict. As a result, their perspectives are under-recorded, an issue also identified in other community-based studies in Yarloop [7]. This makes for a difficult task when attempting to gauge community viewpoints on particular problems or events as the question arises: who is the community and how to define it? While differences in perceptions can be expected among members of any community, the transience of life in Yarloop and the disengagement of newcomers with local issues meant that unresolved problems surrounding the refinery were at risk of becoming superseded by demographic changes.

The perceived disengagement of newcomers may have been due to another factor too. According to Alcoa, the number of complaints by community members dropped sharply between 2000 and 2006 [8], which they interpreted as a sign of things improving. However, by 2006 Alcoa owned most of the land in Area A and many properties in Area B. These properties are rented out to tenants who signed lease agreements in which they agreed not to take action against the company.

Excerpt from Alcoa lease agreement:
> *The tenant agrees that it will not make any claim or lodge any formal complaint against Alcoa for loss of quiet enjoyment of the premises as a result of any damage or nuisance arising from or in connection with any noise, odour, dust or pollution, or disturbance generated as a consequence of the business activities of Alcoa.*

The company denies suggestions, however, of gagging its tenants, stating that the lease agreements do not stop them from making a complaint [9]. Nonetheless, Alcoa's dominance in the local property market may help explain the drop in the complaint figures.

LOSS OF FUTURE

Many long-term residents had seen their future in Yarloop. They had invested in their properties and farms with a view to retire in the area or to pass them on to the younger generation. The upheaval from 1997 meant that some residents put their future plans on hold due to the uncertainties surrounding the land management proposal.

> *It's just disappointing how it's turned out. Well, for me, I thought I had my life mapped out. And now I've had the rug pulled from underneath me. Quite*

> *honestly we don't know what the fuck we're going to do. We really don't. We don't know what's going on now. And you basically, as a lot of other people, their life's been put on hold to a large degree. (local farmer)*

Others saw their dreams for their future come to an abrupt end, finding themselves forced to sell their properties and relocate their families.

> *To lose the farm, even though, yes, we got a fair few dollars, no doubt about that, but we lost something that I thought I'd always have. And when that is passed down to the next generation and the next one if they want it. You just couldn't stay. Well you could stay there all right but your life expectancy certainly would have been shortened in my opinion. (former local farmer)*

> *You leave your plans behind. And somebody else has changed those plans for you and [...] I hate that. (former local farmer)*

For many of the residents we spoke to, their properties were integral parts of their life and an expression of their connection to the area. Investments were made, properties upgraded and much work was put into what had promised to be a secure future. This explains why community members responded so strongly to Alcoa's Land Management Plan which was seen to threaten those futures.

> *So we've got a long affinity with the place, and we came here because it was such a nice place, and now our dreams have been absolutely shattered ... (former resident)*

> *What you see here, I designed and built this place. This is where I was going to live. We had designed it in such a way that it was going to be semi-self-sufficient.*

You'd have your vegies, your fruit trees, things like that; that was the idea. I had been working towards that end up until '96 when the Alcoa issue came into play. And the dreams that I had to do with all this – full reticulation, a bore, fencing all the way around – that's died. (former resident)

The loss of dreams and futures, which for many residents were intrinsically linked to their properties, had a devastating impact. The brief account below highlights a resident's difficulties in leaving Yarloop and the dream home behind.

I will never let go ... I just had this urge for quite a few weeks that I wanted to walk up my driveway [once more]. I parked out the front ... [and] I thought, no, I'm gonna do it. I had to do it. I broke out in a sweat. I wanted to walk down my driveway and I wanted to knock on the door and I just wanted to have a look around. Thank goodness nobody was there ... Anyway I walked down the driveway and all our orange trees, beautiful big oranges we had, beautiful orange trees ... they're all pulled down. I nearly choked. I thought, 'Where's my orange trees?' Thirty orange trees gone, just bulldozed down, olive tree, everything is gone. Look on the other side and that's all gone too. So I walked down the side there and it's all just memories, all memories just flashing through my mind and in my heart and I'm thinking, 'Oh God remember when the kids rode their motorbike through the fence there. Oh my God remember this, remember the ball hanging up there, oh remember the tree they stuck a cubby house in.' All gone, it looked dead, it was horrible. It wasn't my home how I had it, our gardens ... we would just be talked about with our gardens and lawns because we were gardening people. It used to be kept

> *immaculate inside and out. It was all renovated inside and out and it was the most beautiful home with a little wood fire in it. Memories when the kids were all sitting around there in winter you know and I'd light the fire in the morning and they all had their feet on the little lid of the stove, all gone. I peered through the windows and the mats I'd bought from Bunbury were all still there and it was just so sad, so, so sad. When I came out I had tears in my eyes and I thought, 'It's getting further and further away from what it ...' I can't explain it, from how it was when we were there to what it is now is so far away from ... I don't know, I can't explain it, it's just horrible, absolutely horrible. (former resident)*

The conflict also drew into question the more immediate future of local farmers. For some, one pressing question related to the continuation of farming in the area and its compatibility with an expanded alumina refinery.

> *With being so-called organic, it's not tenable for us. We get tested, but only for agricultural chemicals. We haven't been tested for heavy metals. But it's, as I said before, with the organic industry, the perception is that we're living in a clean and green environment and it's hard to tell people that you live next door to an alumina refinery, and it's clean and green. They might think you're having a lend of them. So in that respect it's sort of made our business here slowly untenable [...]. (local farmer)*

> *It's not sustainable anymore because I can't claim that I'm producing safe food if I know that I'm being contaminated by outside influences – namely Alcoa. Knowing that we took samples, their samples, here in the paddock [showing] a high reading of diethyl chloride and methylene chloride [...] and knowing that*

> *we get dust off the trains quite regularly and have done for the best part of 20 years with the alumina. (local famer)*

For farmers who contemplated leaving town, the problem arose as to where they could relocate, since adequate replacement properties further afield were in short supply.

> *Where do we relocate to at my age? I had most things set up here for my semi-retirement and for the family to carry on with. Where do I relocate to? Do I want them to find me a property that is down woop woop, further away from my family, away from ... in between towns? Where do I go that's got these things; that allows me to farm and continue where I am? ... I don't feel like going out somewhere and starting farming again to something different if they can't find me something similar. That is a big worry for us isn't it? (local famer)*

> *Yes, we're farming but we're also investing in our asset. I've got no super. This is my super; we've got this and another property but this portion of our property is on the edge of town, a natural river running through it, deep, heavy soil. The clay here is feet down from the surface. If you're on the edge of the hills, as you go out a kilometre you've got 6 or 8 inches of loam and then clay ... Where am I going to find a property like that? It's all that river silt. (local farmer)*

> *I do not want money, and I'll tell you what, for them to replace what I've got here they're not going to get it for the prices that they are valuating other farmers in the area. Not only that, it's got to be in an area where we want to live as well. I want to live in the south-west. I want irrigated land, the same land type that*

> *I've got on the edge of town, subdivision potential and water the way we've got it. I reckon they've got Buckley's chance of finding it. (local farmer)*

LOSS OF SERVICES

In its prime Yarloop was well serviced by local businesses, and long-term residents remembered fondly how many shops Yarloop had and how good they were.

> *Well, we had a nice big post office, nice shops and what else has gone?*

> *You had the old bootmaker shop in the main street; the amount of shops in that town was unreal. (former resident)*

> *[We] used to have a general store. Alongside the log cabin there was a bakery. Just around the corner ... there used to be a butcher shop there. You had the big two-storey hotel there. That had an electrical business there ... You had five or six service stations in Yarloop. (former resident)*

> *You could get most things in Yarloop. The grocery store in Yarloop was fantastic always and the newsagent, yes, they were really good shops ... You could get everything here. (former resident)*

In contrast, people visiting today may agree with residents' comments that 'there's nothing left.'

> *Nothing on the main street. That shop has got nothing. You go to buy a tin of beans, you can't even buy that.*

> *... well it's just almost no town now. I feel it's just a dying town. (former resident)*

The decline in shops and services was in part historical and related to the downturn in the local timber industry. More acutely, Yarloop experienced sharp economic decline, driven by the exodus of residents after 1997 [10]. Some business owners left due to reduced turnover, 'for health reasons or they're just being pushed out basically.'

> *... it started [with] ... the Land Management Program, when my turnover just died. The graph went straight down. Well ... there have been two factors: obviously one from a downturn in turnover. The other one was in conjunction with health problems. (local business owner)*

> *This business runs off people who come back time and time again as opposed to passing trade ... Now there are no customers in town. Well, that customer base decreased. Well, obviously turnover was going to decrease in proportion with that. (local business owner)*

> *The net income of the town is now so low that it's not going to sustain local business, because nobody has any extra money to spend. They're all living, scratching their arses, really. So there's no spare money around, so it won't flourish as a business place. (local business owner)*

Others needed to leave because they could not get their leases renewed when the landlord decided to sell their property to Alcoa.

> *We're being closed down. [The landlord] own[s] the building; we own the lease, and they want to sell to Alcoa. They had a verbal agreement with us to extend our lease, at the start when we first took it on, as long as we looked after it and made the business worthwhile they would extend our lease. Then Alcoa came on the scene six months later and that was the end of that. (local business owner)*

As a result, services were in decline and residents had to travel longer distances to buy groceries and petrol.

Well, I think we have ... to go further afield for our shopping, if you want to do any quantity of shopping, you've got to go to Harvey [or] Waroona.

The local police station, you go there any time of the day or night and you are very lucky to get a police officer there, they usually work from Waroona or Harvey rather than sit here in the police station. The hospital has just closed. So, you say to me, 'What's different about it?' Nothing is the same, it's completely different ... If you want to do banking you go to Harvey or Waroona, there's no bank, there's not even a bank agency ... It's finished, basically.

Other essential services started to be moved out of town. Despite earlier assurances about its long-term future by then health minister Jim McGinty, the Yarloop hospital service was brought to a halt in 2006 and downgraded to a health centre [11]. At the same time, the Department of Education and the WA Police Department sought to relocate staff from Yarloop to the nearby town of Harvey on the grounds of duty-of-care [12; 13], although a direct link between the departments' concerns and the Wagerup refinery were publicly denied [14]. Concurrently, it transpired that the Department for Housing and Works (DHW) required public housing tenants to sign a form acknowledging that they were aware that their dwelling in the Yarloop area may be affected by noise, odour and emissions from the Wagerup refinery [9]. In fact, as early as April 2002, DHW made the decision to refrain from providing housing in Yarloop until clarification could be obtained about Alcoa's refinery emissions [15]. The Shire of Harvey now also requires written acknowledgement

Mourning the loss of the local hospital (Photo by M. Brueckner)

of potential impacts from the refinery by applicants for building approvals in Yarloop. Similarly, in its capacity as landlord in Land Management Areas A and B, Alcoa has required its tenants to do the same. These precautions taken by government departments served to fuel community worries about Alcoa's emissions and the planned refinery expansion.

Alcoa was not blamed directly for this dramatic turn of events, since many of the decisions responsible were made by government and its departments, but the company's approach to land management was nevertheless regarded a driver of this change. Not only did the Land Management Plan trigger the selling of properties in Yarloop, Alcoa refused to purchase businesses as part of its property purchase scheme. This angered local business owners, who were feeling the effects of a shrinking local population.

> *[Alcoa] will not buy a business. So, if they had wanted to keep the town going, why then aren't they buying a*

business and selling it as they do with their houses to somebody else? But they flatly refuse to buy a business because every business that they buy, they close down ... (local business owner)

This is a business, and they don't want to pay for the business. They don't even want to pay to relocate the business. (local business owner)

Alcoa was also only prepared to buy properties within the Land Management Areas A and B, which for larger farming properties meant that only those parts of a farm would be purchased that were within those boundaries. This was untenable for farmers who wanted to sell their entire farm and relocate to another area. One farmer offers a vivid account:

Down there, I know a couple of the landowners that are quite close [to the refinery] and [Alcoa] come around and they want to buy part of your land. 'Here, hang

Alcoa is hurting this business (Photo by M. Brueckner)

on a minute, you can't just buy half a business.' They won't go out and buy the lot; they only want to buy half a business and by the time you do that you're actually going backwards so people have dug in ... 'Take it all or nothing. If you're going to extend, take it all or nothing.' 'Oh, but this doesn't fall in the A zone or the B zone.' 'Shove the B & A zone up your arse. Either you've got a problem or you haven't. You just can't take half a fuckin' property,' you know what I mean.

FINANCIAL IMPACTS

The questions surrounding finances and compensation are among the most contested aspects of the Wagerup conflict. Money alone was not on the residents' agenda when problems in town began to emerge. Many residents did not seek compensation nor did they wish to be bought out. They had no desire to relocate but instead wanted 'Alcoa to fix its problems' with the refinery.

It was never an issue of selling up and moving out of town; it was for Alcoa to fix up the problem on their side of the fence. And we stated that very publicly ... that we want you to fix it; we don't want to move away; we want you to fix the problem.

The majority, or all the people I know of, actually, [said], 'Fix your problem. Leave the town alone.'

Certainly, some residents in Area A were relatively quick to sell and arguably 'grabbed the money and ran'.

Once Alcoa came along, a lot of those guys all sold up, they got a fair few dollars and most of them had never had too many dollars, money was always fairly tight. Now here they are, they give them a big heap of money ...

> *It was a golden opportunity to go because ... at the end of the day when you fight that long you realise that it isn't going to get better and you aren't going to beat them and there's a way out – you see that opportunity and you go for it. (former resident)*

> *When they introduced the buffer and people started selling their houses that's when it all happened. People thought this is an easy way to get out. The ones that wanted to go, that didn't really want to stay there.*

> *All these people were just taking up Alcoa's offers and they were just running away; they were getting their money and they were going and the houses were just getting left ...*

Many of the properties in Area A were valued generously, providing a financial incentive for residents to sell. Purportedly, however, Alcoa's buyout offers were initially relatively low, and residents refused to sell until the offer was subsequently increased.

> *Area A was looked after, as much as they were prepared to buy their properties at market value, plus 15 per cent, plus relocation cost. There were only two people that took up that offer, at that point. After that, they upped the ante and they decided to make the A area market value, plus 35 per cent, plus $7000, and a lot of people took that up. (former resident)*

Alcoa was known to negotiate with landowners in Area A individually until they agreed to sell. Residents saw this as a deliberate tactic and doubted that a uniform buyout formula was used.

> *It was difficult to maintain community spirit in that sense, like let's stick together and fight this together, and for those who perhaps panicked and just wanted a*

buyout and to leave as quickly as they could. I suppose the community strength had started to deteriorate and I'd say that is what Alcoa was hoping, too, to try to break it down.

If they're after someone they'll get somebody. It's the same with [our neighbour], for instance. When he decided to sell his house it was sold within a month to six weeks and they paid for it ... [He] got everything [he] wanted plus more. There were other people there that were waiting 12 and 18 months, two years to get their house purchased from Alcoa.

They pick us off one by one.

Individual negotiations also had the effect of fuelling animosity among community members as 'some people were seen to be favoured in terms of the buyout process over others and in a tight community like that information spread [...] pretty quickly.'

Alcoa seem to have one rule for some people and another rule for others.

The majority of residents sold only reluctantly. Even though many residents did sell to Alcoa eventually, there was much concern about the associated financial fallout. Many of those who needed to leave town for health reasons, especially in Area B, which offered less generous buyout terms, found it difficult to find a comparable property with the amount they received from Alcoa.

Even the money for our land may have bought us a house block, but would never have bought a house ... They only paid us $70,000, the value put on two acres of land. That's what you are looking at. You are looking at half a million dollars for your two acres of land, and you get bloody $70,000. These are the sort *of things and you are pushed into it and you've got to accept it because your health problems tell you you've got to do it. You can't sit there and fight them ... You're not going to win anyway. (former resident)*

In fact, the property prices offered by Alcoa in Area B were considered anything but generous. Real estate prices in WA had increased dramatically [16], and this was not reflected in Yarloop [17].

The compensation was rats; like Alcoa's 35 per cent of market value. When you can't sell it, there is no market value. For example, [the guy down the road had] a four-bedroom, two-bathroom, double garage, double shed, extensive paving and gardens. They offered him $110,000 for it! You know, you couldn't even buy a block for $110,000. Market value plus 35 per cent. That $110,000 included his 35 per cent. Bloody rubbish!

So the best thing Mum and Dad can do is move away. But even they can't move because they can't get enough for their house because they are in B zone and they're not getting paid enough to actually replace what they have got and all they want Alcoa to do is replace what they have got and get them out of the area. (former resident)

The 'Yarloop factor' – being close to the Wagerup refinery – was believed to suppress local property values.

... it has affected the stigma of Yarloop, it has affected the land values ... People don't want to go into somewhere where they've heard health reports and now they've identified Cookernup more recently as being an area. It again will affect ... values here. (Cookernup resident)

Now I recently spoke to an agent who's got some of these houses ... He's been here before and we were talking about my prospects of sale the other day and I said to him, 'Look, you've got these other places, they're quite nice and they're really relatively priced,' and he said, mate, people ring up and we tell them we've got a house for $250,000, three bedrooms, two toilets whatever, half-acre block and they're like 'how much?' and then they find out it's in Yarloop and then they say 'no thanks' and they don't even come to look at them and he confirmed that for me. He said we have that every day because a $245,000 or $250,000 house that they're describing are unheard of in this region yet we're in a boom, property sales have gone through the roof. These still represent value for this region and you can't sell them but 'there's nothing wrong here' though! (Cookernup resident)

Financial impact (Photo by H. Seiver)

Alcoa was seen to be directly responsible for low property values in the area. The health concerns that had received much coverage in the press, were seen to have affected real estate prices.

They had two valuers: one was Alcoa's and the other one the person who wanted to sell could get a valuer to come in. Now those valuers had already been briefed by Alcoa, even before this thing took off. It had to be within 10 per cent ... I've got reports there, evaluations, that when Alcoa did their value on it and measured it, they left out parts of the property, they under-measured, they used stumps when they had concrete flooring ... they used just about everything you could think of. In one report alone, there was a 98 per cent error on their part.

They ... would bring down the little people and frustrate them to the point that they'd throw their hands in the air and just walk. Because there'd be in-fighting between husband and wife, there'd be a fear factor that the kids are gonna get sick. Families would actually run out when there was a plume and grab their kids and take them inside, and shut all the doors and windows ... And that's been their attitude during the whole thing. That's why you've got your divide and conquer thing ... your frustration ... your valuation ... you have your fear factors which was brought into this. People not sure what was going to happen ... the uncertainty ... they created all this.

For many residents therefore, especially in Area B, relocating from Yarloop to another town came at large financial expense, affecting most severely those living on pensions or those close to retirement, for it meant having to take out a new mortgage.

> *Well, long term what it means for me is that I'll have to go into debt, instead of which I was looking forward to retiring at 60 years of age, I'll probably work until 65 or 70 to pay for the mortgage and dip into my superannuation to pay off my mortgage, where it should be spent on our retirement. So that's what it means. It means extra years working and less money when I retire because I'm mainly paying out the mortgage instead of spending that money on a holiday or a new vehicle or a caravan or whatever, depending on how much I owe, I've got to pay off my frigging mortgage.*

Area B residents on low incomes or pensions felt trapped in that the sale of their property did not enable them to move elsewhere. While they wanted to leave the area, the lack of financial resources and having to cover relocation costs, stamp duty and other expenses meant that they were stuck in Yarloop.

> *... now if I want to move out of here and say go to Bunbury or Mandurah, there is no way that I can do it. So I am stuck here. Whether I am sick or not sick that is irrelevant, simply because I can't get a loan. The other factor is that house prices in Bunbury, or even in Harvey for that matter, are around about $400,000 and up. The only way that I can get out of here is to pitch up a tent somewhere and stay in a tent. I couldn't get a loan ... Yes, I would love to get out, take Dad out and say look I've got a house in Bunbury, but you can't do anything, you're trapped, you just have to stay here.*

For farmers, relocation proved particularly difficult because of Alcoa's refusal to buy farm properties that extended beyond the boundaries of the Land Management

Areas and the difficulty in finding 'prime agricultural land' elsewhere at a similar price. Also, the state government's Supplementary Property Purchase Program (SPPP) did not address the calls from farmers for total relocation.

> *... part of the farm is in [Area] A. They have offered to buy to ... [Area B]. But they will not move past [Area B]. Now with this new scheme the government will buy, well, Alcoa will buy through the government through this supplementary thing, but it still doesn't satisfy our requirements and our requirements are total relocation. (local farmer)*

INSULT TO INJURY: THE 'COMMUNITY ENGAGEMENT' PROCESS

So far, we have focused on the direct effects of the Wagerup refinery on residents. But the manner in which events evolved may help explain how community members responded. A vulnerable community was agitated further by what was perceived as a cold and remote process of community engagement by the company and the government (the role of government is treated in Chapter 5). A different approach by Alcoa might have helped defuse the situation locally and would have been more likely to result in conflict resolution.

Some of these process-related aspects are treated in later chapters. Here, to complete the description of the Wagerup conflict from the community perspective, we provide an overview of the issues under three headings. This can also be understood as a 'wish list' of how the community would have liked to be engaged in the process, with honesty and accountability, respect and compassion, and reciprocal relations.

1. Honesty and accountability

Central to the dynamics of the Wagerup conflict was the question of how the community felt treated during its negotiations with its corporate neighbour. The issues at stake made for an emotionally charged process, which was further compounded by a widespread distrust in the company's dealings with the community from the late 1990s. Overall, there was a widely held perception that Alcoa did not engage with the community in good faith. People saw Alcoa as ignorant of community concerns and arrogant in their treatment of community members.

I think basically they've got to be honest with the community and I don't think they have been. I think a lot of people would back me up on that point.

I think Alcoa has to be honest with us for a starter. That would help.

The notion of dishonesty chiefly refers to Alcoa's insistent refusal to admit to problems with the refinery. When health symptoms first started to emerge among community members, the company denied any responsibility for ill-health or for possibility of any connection between refinery emissions and the impacts experienced in the community. Residents felt that the company was 'in denial'.

They've never admitted there is a problem and they probably never will so they're still in denial.

Number one ... is to acknowledge the problem exists rather than just flatly deny and continue to claim, 'We operate the world's best practice, we are a benchmark.'

They're in denial that they're actually causing any problems out here.

[A] company [... that] will not admit to a problem [is] no different to an alcoholic. If he doesn't admit he's got a problem, they can't start to fix it, and that's exactly what the company's doing. There is no problem, we

> *don't have to fix anything. (former resident)*

Alcoa's reluctance to engage with the community on the alleged relationship between health and refinery emissions was considered offensive, especially to residents who felt they were living proof of there being a problem.

> *... if there's emissions going up your nose and making your nose bleed and making you sick, you can't stop that. That's what they have got to worry about and they won't do that because they won't admit there is a problem. To them there is no problem. (local farmer)*

> *That's all we really wanted from the start was this honest, up-front dealing with the situation. The continual denial is ridiculous. We know there were problems here. We knew a little bit about how the air moved here and how the impacts were increased on us but it's difficult to get through.*

Also offensive was the incongruence between corporate rhetoric about Yarloop's bright and sustainable future and the very visible decline of the town.

> *It's so insulting, all this shit they just ... say there's nothing wrong, you know, things closing down. There's no problem but the teachers don't live here anymore. There's no problem but the coppers have left. There's no problem ... they all say we're going to keep all these things going because that's part of Wagerup's commitment to the community.*

> *Why are the schoolteachers moved out? There are no schoolteachers in Yarloop. They're trying to move the police out of Yarloop. It's not safe to live here ... Those departments don't allow their staff to live here. New Homeswest tenants are given notices to sign that they understand there may be problems associated with*

the refinery, and yet there are no problems.

We had borrowed a video to do something and we were outside and next minute we look up at Alcoa and the plume was black, it was black, black, honestly it was like that. It was black and I'm thinking, 'Oh my God,' so I got on the phone. I rang them up, they came over and I had an argument with [one of them who] stood there and told me it's not black. I said it is black. And we had another guy ... saying it was black and we had a big fight. It was black and he still denied it was black ... I'm thinking, 'But there's four of us looking at it, mate, it's black,' and he said, 'But it's not black.'

We took photos of the [plume] ... we took films of this from different angles. And then their scientists, the Alcoa scientists, they went to view the video. They had a letter written, drawn up, even before they'd seen the video, which stated that is wasn't dark smoke, it was white. [We] showed that to the CSIRO people and they said it was definitely dark. Now these are the sort of things that we're dealing with [with] these people. They deny everything, and make white look black and black look white.

And this is the whole thing, the way these people have operated. Now, how can you have a letter stating that there's nothing wrong, that it's white smoke, even before you see the video. Now what does that tell you? So, now you can understand why no one believes Alcoa. (former resident)

Local perceptions of corporate dishonesty and lack of accountability were coupled with perceived inconsistencies in Alcoa's actions. After consistently refusing to accept the existence of problems caused by the refinery, Alcoa was seen to declare that those very problems had been addressed and rectified. This fuelled cynicism towards the truthfulness of the company's claims.

> *They've always said that there wasn't a problem and eventually turned around and said we'll fix that problem that we didn't have and the problems continued and it's been like that all the time. Complete denial and then we fixed it and then denial and then we fixed it. I think they need to talk to us honestly.*

> *... so originally they denied that we were being impacted and [then] they can go 'whoops, sorry' like you see in some of the press releases. They'll say, 'We didn't realise that it was going to have such an impact,' either socially or whatever, or, 'We found out that the liquor burner was causing a problem for employees and nearby residents but we fixed that problem.' You know that's all part of the PR, that's all part of the feel-good stuff and they just keep pushing that.*

The company seemed to meet community concerns with aggression. In fact, corporate communications were seen to disown the community of the problems experienced locally and to turn on its head the question of 'who is impacting on whom?' Residents felt that they we were being vilified and portrayed as the aggressors in a conflict in which Alcoa presented itself as the victim, unfairly targeted by a hostile minority group in the community.

> *They used to use it all the time ... they call them ... a certain group that just can't be pleased and are only out – no matter what Alcoa does – [they] are only out to hurt Alcoa and prove that they're doing the wrong things, when all the money that they've spent and all the testing they've done has not only been proven once, but many times, that there is absolutely no*

concerns for anyone [...]; and if there was [a problem], they would be the first people to admit it and fix it. If anyone can show them where these problems are, or where they are coming from, to let them know and they will fix them. Sure, this minority of people [is] going around causing trouble. Of course Alcoa are a squeaky clean environmentally friendly company!

People are just out to get money. People are just wanting more because they're greedy, that's how they think. (former resident)

Oh, there can't be any problem there, even though it's been on the TV, radio and in the papers for six years. 'It can't be like that, these people are exaggerating,' and then you've got the government and Alcoa in there saying 'they're all exaggerating, they want our money' so you can never get anywhere because it's the same even at the refinery level.

[They] make themselves look right and us look wrong.

They still complain we're trying to make a political issue out of it. If that's not being in denial, I don't know what it is. They're supposed to put health and safety first.

Many residents believed there was no reason for the conflict to escalate. In fact, it was argued that the conflict could have been avoided and differences reconciled if Alcoa had simply been more proactive in dealing with problems when they first emerged. Community members kept reiterating that 'they did not want Alcoa to stop production; they just wanted them to be accountable.'

If they had of been straight out, open and honest and ... even just a little bit generous, people wouldn't have been worried. They would have even sat down at the

> *table to work out how to stay in Yarloop rather than move out of Yarloop.*
>
> *I'm sure there are dozens, hundreds, thousands – I don't know – companies that work alongside communities. It's all done ... I mean companies are made up of human beings and human beings can solve most problems in the world. I think if you have those sorts of people working together, rather than against each other I'm sure you can solve most things.*
>
> *But, this is our private lives, what we had to do ... all of that, had to contend with, and something that Alcoa has brought on this community, which wasn't necessary in any way, shape or form, 'cause all they had to do was be honest with the community; 'We have a problem here, this is what we can do, I don't think we're able to fix it, but we may be able to reduce things. But we're prepared to do this, this and this for you.' That's all they had to say, and the community would have accepted that, without any hassles.*

Instead of engaging with community grievances the company was seen to have employed a strategy of 'divide and conquer'. Not only was this considered a deliberate corporate tactic but very much a sign of the company not caring about people.

> *Either we're honest and up-front and concerned about people or we don't give two hoots! It's a big problem; it's huge.*
>
> *Alcoa is not accountable for their actions. They have chosen to ignore anything that is detrimental to Alcoa and they started off on the wrong foot because they tried to bounce the people in Yarloop. They came in and tried to treat them like a mob of country hicks and never really sat down and spoke to people and*

listened to them.

For many residents, the acrimony and distrust were such that the conflict was deemed to be beyond resolution. They saw the company's failure to engage honestly with them as a missed opportunity, believing that, after more than ten years, it was now too late for this conflict to be resolved.

> *It's too late now, too late, everyone has gone now, see, if it is well done from the beginning ... before people started to move ... That's the problem now, you see. Who cares? Nobody cares anymore, they don't care. If this was done from the beginning it might have been able to save properties.*

> *I don't believe we can coexist. Not while they're in denial about it so much. If they were to admit there's an emissions issue, but by denying all the time, you feel that they're just trying to avoid doing anything about it.*

2. Respect and compassion

Beyond allegations of dishonesty and lack of accountability, community agitation also spiralled because of negative interpersonal experiences between community members and Alcoa. A crucial omission, according to residents, was an element of respect and compassion, which would have helped defuse the situation. Alcoa was seen to avoid the community, refusing to engage with the issues that mattered to residents.

> *That's what it's been like ever since and they still choose to ignore – to talk – to people. They don't want to talk to people. Alcoa won't talk to the community.*

> *Our dialogue with Alcoa has always been open. We're prepared to talk and meet with Alcoa at any time.*

They've chosen to ignore us.

On occasion, when residents were asked to offer input into company decision-making, many found that their views were sidelined or ignored. The example below describes Alcoa's community consultation process prior to the release of the company's Land Management Plan.

If you have a look at the original consultation process for the buffer expansion, they came out, they did exactly what you just said: told us that was what they were doing. 'We have an issue with noise though only.' [... A] lot of people knew they had a lot of other issues other than that, but okay, based on the noise, 'What are the issues?' 'Well, we have to expand; we want to know where you think we should put the buffer zone.' More than 70 per cent of people said: 'Do not put it through the middle of the town. You can put it either that side of the town or you put it that side of the town and everybody is in it; do not put it through the middle.' So, despite having that recorded [...], they turned around and said, 'We're putting the buffer zone through the middle of the town because that is what the community told us they wanted'! Yet their survey results [...] say [...], 70 per cent of the people did not want part of the town in a buffer; it was either all in or all out.

There was also the question of Alcoa's approach. Company staff were considered by some to be cold and lacking empathy, unable to relate to the human element.

I think Alcoa should have shown a lot more care, a lot more feeling, understood and listened, not just sat there with their tie on and their book and their little bag and treated people like delinquents. They didn't care, they don't care and they never will care. Alcoa don't care about the people, they don't care about Yarloop. Yarloop is just in the way.

I'm just very disappointed by the whole procedure, I think it could have been managed a great deal better. I think there could have been a hell of a lot more compassion on Alcoa's part. (former resident)

Many believed that the company's 'cold and factual façade' was deliberate, designed to confront an emotionally charged community with corporate professionalism.

To them it's money, it's a job, they don't want to know about your life, your health, your anything, they don't want to know about that because they might start feeling human if they put that wall down, they might start feeling for people, but they don't want to do that because if they do that their job's over so they've got to be this hardcore person that is straight down the line, programmed just to do what they are doing and that's it, don't get involved, that's what it is. (former resident)

Alcoa, they're not compassionate enough to even think that's what people do ... [...] [T]hey were like machines; they had a job to do. For them to get close to people was a no-no. As soon as Alcoa would get close to a person [...], they'd be transferred somewhere else and someone else would come in on the scene and that's what it was, all the way through, because they weren't allowed to get emotionally involved with you.

It was straight down the line, it's a job they do. They are programmed to do that job. They don't let anything that is sad affect them. Yeah, it's just a job to them, they're machines. (former resident)

If they were humans with a heart and they were

> *compassionate ... but to them it's a job, it's money and money speaks all languages ... we haven't got a hope in hell, not a hope in hell. (former resident)*

During private negotiations with health-affected residents who sought relocation, Alcoa staff were considered by some to be rude and unamenable.

> *I mean, it's like I didn't exist. The whole lot, it was just like you didn't exist. You were just a pain in the arse, just go away. Get out. People used to say to me, 'If you don't like it, fuck off.'*

> *Of course, a little bit of care, which you didn't get, it was straight down the line. If you want to go, go, there's your 35 per cent blah, blah, go. Not even when we sat there and my husband was nearly crying and just telling them that we don't want to leave here, fix your problem, but that's another problem on its own, it's still a problem and we've been in here for three years. I am angry, I'm still angry and I will probably die angry. (former resident)*

> *We got her to come to our house because we were so distraught we just needed help. We got nothing from her. I laid out everything of my medication ... on the table, I said, 'Help us; you've got to help us.' Help us was: 'You'll get your 35 per cent, that's it.' I hope I never see her again. (former resident)*

In negotiations these problems were complicated by the complex chemistry of alumina production. Non-experts could not determine the impact potential of emissions pollutants – even the government lacked the expertise to deal with this complexity – and Alcoa was seen to use science as a weapon to legitimate its processes and to silence a town and its people who felt intimidated and patronised by the company's superiority on questions of science and engineering.

> They would call a public meeting in the town and they would stand out in the front ... talking a language nobody understood and they would talk for 10–15 minutes showing graphs and bloody whatever, and then they would say, 'Any questions?'

> You've been given written information and you can't understand it, it's all in figures and you know that they're doing that because you can't understand it. It was just absolutely dreadful.

> They had their fingers on the pulse and [were] bamboozling everybody with a lot of long words and technicalities and all this. It was just all information and data that they brought in and no residents were allowed to get up and have their say.

> I also get the feeling with the dealings I have had with them over the years that we were just Yarloop Yocals, just semi-literate, not much between the ears, not much brain matter, so they could pretty well pull the wool over our eyes and blind us with science and facts and figures and they send in some people that are quite, you know, all the suits kind of thing and all the very super slick sales type of people, you know, academic, very, I can't think of the words right now but they are the sort of people that kind of make you feel lesser because that is what they are skilled at doing.

In some residents' eyes, Alcoa exploited the expert/non-expert dichotomy created, which served to insult their intelligence and discredit local knowledge.

> They don't understand even in a 'simple town' like Yarloop, I use that terminology, people have connections and the 'simple' guy on the street isn't

> *as simple as everybody thinks. They might not have a uni degree and they might not know this and that but they've lived and they've had their parents pass on information and they understand things in a far better way than a lot of people ever think.*
>
> *That's how I feel and they treat the people of Yarloop like the peasants, like they can't think for themselves or be sensible, logical [...]. We can do what we like with them sort of thing, like puppets on a string, which is jolly scary and really sad. (former resident)*
>
> *We're just little nothings to them. (former resident)*
>
> *Well, they shouldn't have come in and treated the locals originally like hillbillies, country hicks.*

The community's non-expert status was exacerbated by it being deprived of access to information. Community members experienced great difficulties when trying to access company-related information such as emissions data and health reports, which would have shed light on the community impacts the refinery emissions were believed to be causing.

> *When the community requested to see various reports and wanted to speak to the analysts at the Chemcentre where Alcoa was sending their samples [... under] freedom of information, the hoops that the community had to go through to try and get the information when it was available, it was just ludicrous.*

In the absence of available data, community members had to resort to obtaining information through sources that wanted to remain anonymous, creating difficulties in proving to Alcoa and the government that their data had integrity and was valid.

To add insult to injury, residents felt mocked by the company when their concerns and grievances were reportedly met with disregard and their state of health disbelieved:

> *If I was to go there and say, 'I'm sick, I need compensation,' they would just think 'another fruitcake', basically. (former resident)*

> *They didn't believe me with all my medication. They didn't believe me that I had lung functions done and the first one was bad, the second one was worse and so on. They didn't want to know about that. They didn't want to know about the antidepressants, they didn't want to know about anything. ... I didn't expect them to side with me but to understand, and they didn't, they didn't want to know about it. (former resident)*

To many residents, Alcoa was nothing more than a profit-motivated multinational, prepared to sacrifice a community to maximise revenue. They were convinced that the company did not care about the community and that its philanthropic gestures were strategic and tokenistic.

> *They are not interested in people who don't make money for them. We weren't making a dollar for them, so we weren't important at all. It's only the dollar that counts, obviously, because they're accountable to their companies in America for making the dollars. So as long as you look after the people that are going to make the dollar for you, you are fine. Alcoa is a big multinational company. (former resident)*

> *No compassion at all. They can say they do; they flick out their wallet, 'Here, we'll put a playground up in Yarloop, oh we'll do this for Yarloop, oh we'll do that for Yarloop.' There is no Yarloop. 'Yarloop is a booming town.' It's not a booming town, it's a dead town; it's a dying town.*

Now this is the sort of people we're dealing with. They don't give a damn. The only thing they're interested in is the bottom line ... their bottom line, which goes back to the shareholders in the US, and nothing else. They kick the little guys when they're down, and there's no compassion at all there. And I think that anyone that does seem to show any compassion there ends up leaving the company.

These stories highlight the serious power differentials at play. Community members had no means to hold the company to account. While the onus of proof rested with the community – it needed to demonstrate that it was adversely affected by the refinery – local knowledge was discounted by a corporate science which supported the company's claims of being a benign operator. Allegations made against the company were dismissed as being emotive and unscientific. Alcoa was in control of the process, determining its direction and outcomes, while residents lacked the means of countering the company's science and engineering prowess, and ultimately the means to defend itself.

3. Reciprocal relations

While Alcoa and the state government made attempts at consulting with the community in an effort to reduce the friction, the various consultative mechanisms employed were widely regarded as flawed and tokenistic [18] and biased in favour of Alcoa in terms of member composition, agenda and process protocols. They were regarded as 'rubber stamping committees' for the purpose of legitimating the company's courses of action.

They ... choose pro-Alcoa people for the committees, people who will just follow whatever data the company throw at them ... question it as little as possible and put a positive spin towards the company on things, rather than use any information to try and make things better.

They get every increase they want in production. Supposedly there's community input into the licensing but really it's null and void. The community has no voice and the machine just rolls on. It's out of control as far as I can tell.

I think a lot of that community consultation has been orchestrated.

The consultative committees were also considered highly unrepresentative, leaving out voices that were critical of Alcoa's operation.

They don't treat it as a community. They pick out some people, and that's the community. And those people they pick out, have a say on what the community requires, and the community is not consulted. The decisions are being made supposedly for the community by unrepresentative bodies.

[I] tell you, the infiltration or the 'white-anting', it is unbelievable ... when we talk about a community, say you have a consultation, I don't believe in consultations anymore, a community cannot – here anyway – cannot self-determine ... Alcoa makes this an artificial community; it's an artificial community and it is a community that is being used, abused, stripped.

The other key criticism was levelled at the limited decision-making powers of the various committees. Even though residents were assured that the consultation process would enable the community to have input in local decision-making [19], residents felt that outcomes were predetermined and that they were simply being told what was going to happen.

> *They think community consultation is to send out colour pamphlets to everybody, which have gotten now to number 22 in the series. And after number 5, they go in the bin, don't they. But then they can say, 'We told everybody.'*

> *... they would walk into the committee meeting and say, 'This is what we are going to achieve today. We are going to achieve an agreement on the expansion or this sort of thing,' and to me that is not what I consider community consultation.*

> *It was always what Alcoa wanted, how they wanted and when they wanted [it], regardless of what anybody else thought, wanted, desired or anything else, and that's how the process has been.*

> *I wish to goodness I had never heard of it ... you can't have a local agenda, what the people want. You go through all this sitting at round tables with butcher's paper and pen and ... all those other people have already worked out what is going to come out of this.*

Overall, local experiences of community consultation seemed to confirm suspicions of dishonesty and fuel the belief that Alcoa was intent on destroying the town. All of the company's actions were seen to demonstrate that Yarloop was in the way of Alcoa, and the town's demise was a matter of strategy and only a question of time.

> *It's all politics; they'll keep out of it and let all their little runners do all the talking and everything else and they will just sit up in their little high horses up there and think, 'It will all go away one day.' It will because people will get so old and so sick they won't be bothered complaining anymore. Old people have dropped down dead.*

> *I said [at the meeting], 'I believe, and other people have said to me, we should have a general meeting of Working Group members so that each Working Group understands what the other ones are doing and people have the opportunity to ask questions across the groups and get answers and so on.' [They said] 'That's a good idea, we'll do that later.' I said, 'People are asking and I think we should organise it now, could I have the contact details for other members of the Working Groups?' 'We're not at liberty to divulge that information, but we do think it's a good idea and we will do it.' They did do it after all the consultations were completed; they had a wrap-up meeting to bring all the Working Groups together and have a few sandwiches and a couple of drinks and pat everyone on the back but with no opportunity for cross-interaction between the groups other than individual people that happened to know other people who asked questions.*

These views were echoed more widely; a member of state parliament summarised the consultation process as follows:

> *... the community has been expected to give up a lot of time to negotiate and to be involved in conflict resolution processes effectively, but they have always been controlled by the company to ensure that the outcomes were such that they never resolved the conflict, because to resolve the conflict would involve the company in doing something it didn't want to do and that was to reduce its air emission impacts by reducing its production.*

The prevailing sense was that the community had given up a lot for very little in return.

CHAPTER 4
THE CORPORATION'S PUBLIC STORY

As part of the research conducted by Edith Cowan University between 2002 and 2003, our contact with Alcoa personnel left us with an abiding sense of a company feeling like it was under siege from segments of the local towns. Try as they might, Alcoa could not get angry residents to agree with how the company saw things and could not get them to be quiet in the media. The company's property purchases subsequently played a big part in getting many of the most vocal locals to go away, although Alcoa stated publicly that this had not been their intention.

The opportunity to witness the internal functioning of Alcoa Wagerup at a critical point in its history, when things were unravelling with local communities at an alarming pace, was a privilege. Alcoa was implementing its land management strategy amid high-profile national media attention. The headline 'Uncle Al Stinks' [1] in a national newspaper in early 2002 brought Alcoa to its feet; it needed to stop the problem getting worse.

But then, as now, how the problem was defined proved to be a strongly contested area, although it was always Alcoa who controlled the definitions of what counted as the problem, as well as responses to it [2].

IS ALCOA HURTING TOO?

We observed many key personnel struggling with feelings of confusion, anger, frustration and hurt, related not only to how some people in the towns were treating them but also to the pressure on the company to account for itself, particularly to the Parliamentary Inquiry [3]. Alcoa managers expressed strong pride in their company as a leading light in the resources sector in terms of the environmental sustainability of its mining and refining operations. Many in senior leadership positions believed Alcoa was a good neighbour and they could not grasp how people could be blaming it for problems it felt were of no substance. Whether these beliefs were deluded or not, there was no doubting the genuineness of their feelings.

These feelings of distress tended to exist in the absence of any critical awareness of the power and privilege Alcoa wielded as a multinational company in WA's pro-development political climate. The personal feelings of Alcoa managers could be interpreted as individual reactions to perceived threats, which then played out in reactions against complainers and in defensive responses to the negative media.

To Alcoa's credit, with the ECU project it was allowing outsiders in, and it genuinely wanted to know what would make things right with the local communities. But while it was soon clear to us that a more expansive response by Alcoa to the criticisms of the townspeople would help make things right, for many reasons (explored in Chapter 6) the up-close involvement of a university research team was unable to shift the defensive reactions and feelings of hurt and frustration of its managers.

For some residents in the Yarloop district it would be quite provocative to read a chapter on how Alcoa has been hurting and the adverse impacts it has suffered. Many locals would say that Alcoa caused the problem and so it serves them right if they are feeling some of the pain. Some did not at the time accept the personal apologies and expressions of concern offered by Wagerup managers [4]. They saw Alcoa's decisions as emanating from distant boardrooms. This placed individual managers in an invidious position as they tried to respond to the evidence of loss and suffering while holding the company line that they were not causing the problem.

There was, in 2002, and possibly still is, very little awareness in the community of how Alcoa's front-line staff experienced this conflict. The progress achieved as part of the research at the time [5] related to the integrity of the relationships built between some of the managers and many of the local townspeople. These relationships existed in a context of unevenly held power in influencing the outcomes [2]. Alcoa had tiers of authority and money and other resources directing the local problem-solving parameters. Residents' awareness of this limited their trust in local attempts to mediate between the company and the community [6].

Some of Alcoa's own factory-floor employees have been very vocal in various public meetings in Yarloop in asking Alcoa managers 'to come out from behind the dollar signs and [for the company to] show its human face' [7, p. 16]. In late 2002 one manager at Wagerup wrote:

The impacts of this project are being dealt with and felt at all levels of the organisation. [It] would be good if the human side to Alcoa could come across [in your book]. My personal view is that as an individual working on this project I am very much affected by everything that is done, said, felt, etc. by Alcoa as well as the community and media. So too are the people from head office who have never even visited Yarloop. All the people from Alcoa dealing with this issue are human and feel very much for the people of Yarloop and the impacts on them ... [we are] humans with feelings working on behalf of Alcoa rather than as a giant corporation that has no feelings.

That Alcoa personnel were distressed and concerned says something of the challenge the conflict came to represent to the internal functioning of Alcoa. The company was and continues to be aware of the conflict, and from its perspective, has worked hard to solve the problems. However, we resist descriptions of Alcoa as if it were a human entity and concur with the argument in the film *The Corporation* [8] that it is dangerous to attribute human qualities and rights to a multinational company whose primary objective is to make money for its shareholders. At the same time, we are not presenting the story of Alcoa as if it is representative of all employees' views, or any one person's responsibility. We believe that the company as a commercial entity needs to be held accountable for harm done and this accountability cannot be on its own terms or be left to the legislative powers of the state government alone.

In this chapter we seek to explore whether, by its own account, Alcoa has accepted that it created many of the parameters of the conflict and whether it has attempted, to any extent, to solve aspects of the problems. The question posed at the beginning of this chapter – is Alcoa hurting too? – thus becomes: has Alcoa, as a company entity, shown evidence of learning from the experiences of the last decade such that it can now reasonably expect to have a community licence to operate [9]? And from a social justice perspective, is Alcoa's community licence to operate based on the informed and willing consent of those people who believe they have been adversely impacted by Alcoa's operations?

NO ACCOUNT IS NEUTRAL

As an account of a serious and ongoing controversy between members of the public and a mining company, it is not possible to write this or any chapter in a neutral manner. We have already declared our intention of giving voice to the community members. This orientation derives from a critical approach [10] to the play of power [11] which has consistently favoured the government and the company at the expense of local people and, less obviously, the rights of the general public.

We try to present an accurate account of the arguments and efforts of Alcoa to respond to community and government concerns. At times we provide counterpoints to Alcoa's claims, where this information is on the public record or was formally accessed through the research for the book. We do this to address the continuing unequal access to space to be heard. A content analysis of Alcoa's claims is also presented in this chapter to show an alternative reading of the 'facts'. This alternative reading is developed in Chapter 6 to show how Alcoa's use of centrally controlled corporate power and claims of scientific knowledge are used to maintain and grow its operations.

No account by Alcoa of the situation, even if based on claims to scientific objectivity, is neutral either [12]. This is particularly so when Alcoa, the company, seeks to maximise its own market competitiveness and when it interacts with local communities from a business vantage point. Science serving the ends of profits for company shareholders is already situated in a particular power orientation of 'us and them' [13].

What counts as a valid knowledge basis to an argument tends to depend upon who is talking, with Alcoa claiming expertise based on scientific and rational knowledge and community members claiming expertise based on lived experience [14; 15]. The controlled speaking position adopted by the company for its public statements about the conflict was perceived by many locals as not lining up with private realities and conversations. We suggest that this split between the public and the private is another site where power is exercised such that Alcoa maintains its position of relative control and influence. We discuss this idea in more depth in Chapter 6.

Some local people became very skilled in accessing and interpreting scientific and technical reports in their efforts to understand and challenge Alcoa's claims. They also made a substantial contribution to the science of air monitoring through the use of the 'bucket brigade' grassroots strategy of gathering independent air quality samples [16]. Nevertheless, little credence was given to arguments on either side, no matter what the source of the knowledge, and conceding to the other on any significant point was rare. This tended to mean the status quo was maintained which already advantaged Alcoa over the local communities.

There is a pattern to the communications from Alcoa during the controversy in which the company was either denying responsibility or defending its position. It has tended to proffer scientific knowledge that denied the company's responsibility for pollution and social impacts, or defended the company's decisions. Despite this, there is evidence of a divergence of views within Alcoa management on the nature of the conflict, particularly relating to Alcoa's handling of its community relationships at critical points. Several Alcoa personnel, and international experts acting for Alcoa, have brought some critique to Alcoa's public relations claims of being a good neighbour and causing no harm.

CORPORATE COMMUNITY RELATIONS

The task of determining what counts as Alcoa's perspective and how to represent it is a challenge, due to the space the company speaks from. When Alcoa issues a media release, its content is not the voice of an individual who can be held personally accountable for their comments. Even if a public relations person is identified, it is common knowledge that they must 'toe the company line' and may in fact hold different personal views on the topic in question. In speaking with one voice the sense is conveyed that there is one truth and all Alcoa personnel, including workers who live in the adjacent communities, concur with the line of argument. But occasional inconsistencies and differing points of view in its publicity unsettle the dominant story-line by 'Alcoa'.

The multilayered nature of the problem, the large number of parties involved and the complex pulls and pushes in positioning in some quarters makes analysing the situation fraught. Our solution is to write about Alcoa in two main ways. Firstly, we speak of the *generalised* Alcoa when to do so does not undermine the reality and politics of a collectivity of human activities that comprise the company. These include historical and documented activities – the extra-local relations of ruling [17]. In this respect we refer to Alcoa as a non-fixed and non-monolithic entity [18] comprised of many truths, tensions, accountabilities and potentials. For example, at times residents felt that Alcoa managers would claim that they were trying to solve certain problems while at the same time they were aggravating other aspects of these problems. The other way in which we write about the company is as a *corporate entity* which places constraints and sometimes contradictory requirements on its personnel.

A particular frustration local community members experienced in attempts to engage Alcoa in dialogue, was that direct contact with national and international company managers was not possible. This became a significant factor when, after months of community-based open meetings in 2002 and 2003 with local Alcoa managers and personnel, community members perceived that a directive from higher up the chain of command closed down negotiations on property buyouts. A further source of frustration for local people was when key personnel moved on, often at short notice, taking with them the goodwill and knowledge of the endeavours to sort through the issues, and the trust that had been developed in particular relationships.

'Alcoa' was always someone not present at the table, something more than the Alcoa personnel who attended the meetings or answered the phones to receive a complaint. Yet for many residents, Alcoa pervaded every aspect of life: their private spaces and matters, and their financial situations. Outspoken locals became very visible and were often judged for standing up against Alcoa, while 'Alcoa' had the capacity to be something other than the people on the ground trying to handle public relations issues. We endeavour then to avoid using the generalised Alcoa as if it acts separately from people in positions of authority in the company.

DIVERGING ALCOAN VIEWS

The following excerpt from the Standing Committee on Environment and Public Affairs [19] provides a detailed account of how then managing director of Alcoa World Alumina in Australia, Wayne Osborn, wanted to portray Alcoa's position.

I want to place on this record Alcoa's unreserved apology for its part in a particularly sad period in an otherwise happy and mutually rewarding 40-year history in Western Australia. Alcoa is committed to restoring the trust and good relations of the community ... The complexity of the issues faced by Alcoa and the community surrounding the Wagerup refinery have been among the most difficult and challenging for any company and perhaps any community in Australia. It has involved a range of different and sometimes interconnected factors and events which have impacted on all of us.

It has involved achieving the necessary reduction in emissions, addressing health concerns, pursuing scientific evidence and greater information, dealing with industrial issues and land purchases and dealing with the community mix and future viability of local townships. All this has been done in the context of deeply felt emotions ...

All alumina refineries have an identifiable odour. What is unique at Wagerup is the lack of an adequate buffer and the close proximity of the refinery to [...] semi-rural lifestyle and residential landholdings, many of which have been built over the last ten years ... The absence of a coherent, formal land-use framework has been a root cause of the problems at Wagerup. Although Alcoa sought to acquire land around the refinery, gradual encroachment of residential development, over which Alcoa had no control, resulted in land uses that rested uncomfortably with refinery operations.

Unacceptable odour and noise from the liquor-burning unit during 1996–7 provoked the breakdown in relations between the refinery and its neighbours. Since then, Alcoa has been able to successfully address the odour and emission issues and it has provided a path forward for those employees with health problems. However the social issues remain.

The land management strategy, which resulted from consultation with the local community, was intended to achieve a resolution. It was meant to allow those who still harboured health concerns to relocate. It achieved this for some. However, the social impact of the controversy on the community, together with the accelerated movement of people in and out of the town, has proved to be a compounding problem rather than the solution it was intended to be. The resulting changes in community structure and composition have impacted the community in ways in which Alcoa could neither have imagined nor never intended. Longstanding residents have seen friends and neighbours leave, businesses close and the social fabric of Yarloop change. It has provided a real sense of dislocation in people's lives. Many Alcoa employees and their families live in Yarloop and they too have been impacted upon by these events, including loss of trust between their community and their employer.

... In this context, Alcoa has always sought to act with the best intentions with the information available to it at the time. With hindsight, our motivation cannot be questioned. Where these decisions and actions have caused distress, I personally, on behalf of Alcoa and individuals within the company, unreservedly apologise. We deeply regret the situation. We have done what we can to redress any adverse impact those decisions have had on people's lives. However, as much as we may wish to, we cannot undo the past.

The central message I would like to leave the committee with today is that Alcoa has learnt from these painful experiences and it is working with the community, the government and planning authorities to move towards a more positive future ... Alcoa now realises that it placed too much emphasis on scientific and technical issues and not enough emphasis on the human aspect or response. People had genuine concerns and they were being impacted upon. With the clarity of hindsight, Alcoa should have immediately shut down the liquor burner. By the time it did so (temporarily) in 1997, the damage to Alcoa's relationship with some if its workforce and the

> *community had already been done. Alcoa intruded on people's daily lives. It stumbled and lost their trust and confidence.*
>
> *... Alcoa has also committed that we would consider expansion to the refinery only when there was broad community and government support. ... Alcoa understands that the Yarloop community needs time to heal and it will take time for us to regain a constructive relationship.*

Mr Osborn attempts to provide to the Parliamentary Inquiry a developed sense of the interrelated issues and, to a naive reader, his claims are plausible and couched in an emotional language of concern for the Yarloop community. He did not, though, seek firsthand accounts of the issues or accept invitations to meet with the people affected. It is hard to enact responsibility on behalf of Alcoa separate from a commitment to, and relationship with, the involved parties. Mr Osborn speaks with authority for Alcoa *and* the Yarloop community and locates this account within a business desire to 'consider expansion to the refinery'.

Thus, at the peak of the controversy, in a formal account of the issues by the most senior Alcoan manager in Australia, the persistence of corporate power in simultaneously achieving its goals and its invisibility is evident. This dynamic is on the public record as fact and not unsettled as a contradictory and possibly irreconcilable conflict of interests. 'A more positive future' is not very likely to arise from this mix of uncritical appreciation of corporate power, insufficiently declared vested interests and a naive (some would say disingenuous) belief that 'Alcoa has learnt from these painful experiences.'

Wayne Osborn retired from his role as managing director of Alcoa World Alumina Australia in early 2008 [20] and now holds a state government appointed position as chair of the state superannuation fund, GESB Mutual.

The following account of the Wagerup conflict offers a point of view from within Alcoa. The interview conveys a different perspective to the majority of public statements and media releases by Alcoa. It appears that internal efforts were made to change the combative approach shaping Alcoa's dealings with the Yarloop community, but with little success. This personal account of the issues begins with comments on the workplace health concerns of a number of employees, which preceded and perhaps precipitated the escalating tensions between Alcoa and the local community in 2001 and 2002.

I don't think the company engaged with the employees in the first instance particularly well. Our view was a very strong-armed view. We took a very legalistic view whilst having an argument with the union. Part of their issue was we didn't recognise that we had an issue so it became more of a philosophical argument than anything else. I think everybody wanted to do the right thing but the science ball said there wasn't a problem. I went to two community meetings down at Yarloop and the first one I'd say there were probably 200 to 300 community members there who really stood up and just said 'I've got this problem'. It was very disturbing. As someone from Alcoa sitting in the audience it was a very disturbing situation but the view still was very heavily that we weren't doing anything wrong. We had all our bases covered, we had all our measurements. We had all of the health guidelines and so forth and there wasn't an issue.

[These internal issues] ... ended up getting into the press. Even though the people were being paid they had significant issues that they didn't believe the company was recognising that they really had an issue and that the company had an issue. It then got

into the West Australian and the whole thing just started to snowball.

... I don't think the solution to this problem is going to be someone suddenly measuring something and saying, 'Eureka, we've found the problem!' ... There are lots of people within the community who don't believe our data so there is all of that argument. My input would be: there isn't any hidden agenda or any hiding of data that I'm aware of or any conspiracy theory. The data's the data. If someone else measures it they'll find the same results. That's not what the issue is about; it's about relationships and a breakdown in relationships.

The reality is, when Mark Cullen [Alcoa's Chief Medical Officer] came out I think his biggest advice to us was to really just apologise, put action plans in place, to make some changes and to let the whole thing calm down. It's a pressure cooker – you needed to let the pressure off. You need to respect the people; you need to build relationships; understand what's going on. I think basically we followed that for a short period of time and then the drive to expand the plant and increase the capacity – improve the financial outcomes – I think overcame all the rest.

The other question which I think was absolutely relevant was 'well, why didn't Alcoa do the health study around the area?' My advice was to do it – and there were good reasons not to do it. Whenever you do a health study you'll find something wrong, so there is a negative which I'm sure if we did find something wrong it would have more to do with lifestyle and other issues than it would about Alcoa, but you'll find something wrong; that's invariably what happens. There is some logic not to do it but I think in the situation we were in if we were looking at trying to rebuild trust and rebuild credibility then we should have been doing it.

I essentially negotiated for us to do it and I got the rug pulled on me on that one as well. My views were somewhat at variance with everyone within the organisation.

There was a difference of views within the organisation and there were decisions made. I'm not trying to be critical of how the decisions were made but I think it's a relevant question: 'why didn't this happen?' But I don't think that we as an organisation were really trusting of the government either so you have a lot of parties involved in this who didn't have the level of trust to enable you to work through the issue effectively.

I think Alcoa's perspective again was the Health Department should come out and say that everything is okay, there isn't a problem; they wouldn't say that so therefore they couldn't be trusted.

Also there's this delicate balance for any organisation to take things seriously, not ignore what's being raised and work the issue because, certainly, in the initial stages what people want is the issue addressed. They don't necessarily want you to wave a magic wand, but what they don't want is being denied and so it's an issue of how can you compassionately deal with people who have certain information, have certain beliefs, how do you effectively deal with that without it actually going into mass hysteria which I think is where it's ended up and it still hasn't been backed off today. I don't know that the issue is any better today than it was five years ago.

I just think the impact and the way we managed through it and so forth was a disaster. I think there were a lot of mistakes made that in fact had unintended

> *consequences but I think if there was a bit more thought put into it those unintended consequences probably would have been foreseen.*
>
> *I think we went down a path that was 'we're going to get our supporters and we're going to battle this out' rather than 'we want to reconcile with the community'.*

Another Alcoa manager held views more in step with the 'company line' and excerpts below reveal the belief that the company is trying to do the right thing by local communities.

> *Alcoa certainly didn't want to see the town die or depopulate; it wanted to see the opposite. It wanted to see the township grow and be prosperous and be a good place to live. So that's what we were trying to achieve with the land management strategy. It's interesting to look at the complaints changes and the drivers because I think it potentially tells you a lot about outrage and how you can impact on people and so on and a big lesson out of this is Alcoa put the land management plan in place because they thought it was a really good thing to do and I literally remember people saying to me at the time that they felt trapped and they had no way to go. I remember Alcoa going and saying that here's what they intend to do, that there was going to be some objectivity on the lines north and south and it would allow people to leave gracefully so it was seen as a good thing. The unintended consequence was it outraged a whole pile of people.*

As far as this manager was concerned, there was support for Alcoa from the communities around Wagerup.

> *I guess my interactions with the locals over the*

last few years have been that the vast majority of people are pretty happy. They don't like the degree of controversy that surrounds the place because it drags their area down, it drags their property down and it can be very divisive in the community. I've seen similar things in rural communities in New South Wales. The majority of people are pretty comfortable and happy to stay there and want the town to go ahead. Some of them get pretty pissed off at some of the very public slanging that goes on because they think of it as them and their community and I think most people definitely want to stay. That's my view.

I also think that those who believe that it's totally incompatible having a community with an expanded Alcoa next door, that they are in an absolute minority within the town.

We do polling down there the same as companies all around Australia and all around the world and that's the sort of stuff we ask. There is very high support for an expansion that manages emissions, sees local businesses benefit, local jobs and sees the townships go ahead. There's very high support for that, very high; well above three-quarters and that's been part of what we're trying to do as well. Managing emissions has been a story for many years now, that's something they've focused on and put a lot of money in. Some of the engagement that we've had with people through the consultation process has been about them trying to help us find ways to deliver the other things; things like the Enterprise Learning Centre, that came out of community consultation down there with people. The Wagerup Sustainability Fund, that's where that came from. That's people saying that they want to see us inject a significant amount of money into the local community. They want to see that grow if the refinery grows and they want to see it spent on township sustainability projects. There's very good support for that which is good which is what Alcoa wants too. On the other side of the coin there are some people who don't want the company to grow or don't want to stay there if the company grows or don't want to stay there anyway and both the Area A and B policy and now the thing the Government has just done gives them that opportunity. That won't make everyone happy but it does at least give people a chance if they feel trapped they're able to move.

With regard to the continuing disquiet in some quarters, I think it needs constant attention, I'm certain it does. If Alcoa stops listening then that thing (consultations about the expansion) will go in the wrong direction without a doubt. If it doesn't deliver on its commitments and it doesn't deliver on local benefits then the things that people expect those local areas should be able to get from having a major mineral company nearby then that will change, people will become increasingly unhappy if we don't do those things so it requires constant attention.

ALCOA'S PUBLIC COMMUNICATIONS

Risk for Alcoa came to be located as much in its communications with local communities and the media as with its refinery emissions. Here we present a selection of Alcoa's communications in the public domain to show how the company sought to present itself. The company was able to speak from a range of spaces, including its submission to the state's infrastructure strategy; media releases; its own website and research; and meeting notes in a local newspaper from an Alcoa-initiated consultative network. While the selection does not provide a full account of the material in the public domain, it does convey the main points of Alcoa's arguments.

The company initiated much of its own publicity and research, and it was also often asked to comment on press releases presenting community members' concerns. We include here some examples of press releases, with one excerpt delivering a vigorous critique of Alcoa from the highest authorities in the legal system. Chapter 6 revisits many of these claims by Alcoa through the lens of the land management issue.

Alcoa's mission statement according to the company's website in 2002 [21]:

At Alcoa, our vision is to be the best company in the world – in the eyes of our customers, suppliers, communities and people. We expect and demand the best we have to offer by always keeping our values top of mind.

Alcoa's values, as summarised in its 2008 sustainability report [22]:

Integrity *Alcoa's foundation is our integrity. We are open, honest and trustworthy in dealing with customers, suppliers, co-workers, shareholders and the communities where we have an impact.*

Environment, health and safety *We work safely in a manner that protects and promotes the health and wellbeing of the individual and the environment.*

Customers *We support our customers' success by creating exceptional value through innovative product and service solutions.*

Excellence *We relentlessly pursue excellence in everything we do, every day.*

People *We work in an inclusive environment that embraces change, new ideas, respect for the individual and equal opportunity to succeed.*

Profitability *We earn sustainable financial results that enable profitable growth and superior shareholder value.*

Accountability *We are accountable – individually and in teams – for our behaviours, actions and results.*

The company's values are also expressed in their 2001 Land Management Policy [23].

We know people want to live their lives without worrying about their health, their property values, or emissions from Wagerup Refinery. We know people want their community to stay together, and we know people expect Alcoa to further reduce emissions. Alcoa remains committed to:
- *Reducing odour and other emissions,*
- *Reducing noise,*
- *Investigating health concerns.*

Alcoa also remains committed to:
- *Protecting property values,*
- *Supporting the integral nature and quality of the community and encouraging people to stay,*
- *Making it easy for those who wish to leave to sell their properties.*

Alcoa will invest in the future of the local communities, and is determined to be a good neighbour, both now and in the future.

As far as local communities were concerned, the company's strategy of buying up private properties in the middle of a conflict about pollution undermined any good intentions it might proclaim.

Alcoa advertisement, *Harvey Reporter*, March 2002

Alcoa's Wagerup Refinery Manager, Ann Whitty, today apologised for a potential breach of the refinery's environmental protection licence on Wednesday night. The refinery exceeded its dust limit ... when dust control equipment on a calciner failed and the unit was not shut down as quickly as required ... Ms Whitty said because of the wind strength and direction at the time of the incident it was believed there was little impact beyond the refinery and the company has not received any external or internal complaints. [24]

Local community members reacted to the language of 'dust' and the term used by managers of 'a dust excursion', claiming it covered up the reality of it being pollution. A subsequent dust excursion was subject to an inquiry by the Department of Environmental Protection. One of the university researchers was in Yarloop on the day and was interviewed.

***Harvey Reporter*, March 2002 (excerpt)**

Alcoa has accepted full responsibility for a complete and effective resolution of environmental issues at its Wagerup refinery. The acceptance has come after a visit by world authority on environmental health issues, Dr Mark Cullen. The Yale University professor who is also Alcoa's chief medical officer and advisor on these matters found that there was a problem at Wagerup and had recommended urgent action. He said complaints about allergic reactions to the refinery's emissions were consistent with other similar plants around the world but said the number of complaints was much higher. [25]

Wagerup Community Consultative Network, September 2002 (excerpt)

In recent times (Alcoa) Wagerup has been working to replace the population in Yarloop, but there has been a huge change in the community in terms of people moving in and out. This has been a huge social change and upheaval so we are working with ECU, because the university is independent, to try to improve the social aspects of this situation. ECU is recommending how Alcoa can do things better and is helping the community to understand how they can also contribute. Alcoa has recognised that we want to do the right thing but we don't have all the expertise to help with the social aspects because we tend to have people that look at things from a technical point of view, not a social one. (Ann Whitty, Wagerup Refinery Manager) [24]

Locally, Alcoa was criticised for engaging and co-opting universities and here we see a glimpse of the politics of how such partnerships can be used in the media. Alcoa, allowing for its lack of expertise in social sustainability, is able to say some very troubling things about changes in a community due to its actions. This is done without conceding ground, due to the association with Edith Cowan University. A local manager told one of the researchers that Alcoa does not have a language for talking about the social aspect of its corporate social responsibility. They compared it to the challenge Alcoa faced twenty years earlier, when they had to address the environmental issues of their operations.

Wagerup Community Consultative Network, February 2003 (excerpt)

Late last week the Kwinana Refinery notified the

DEP of residue dust data that had been altered by a residue employee ... A number of corrective actions are being implemented to prevent this situation recurring. These include upgraded controls on computer program access, additional dust analysis and random monthly audits. Alcoa will continue to consult with its neighbours and the community to keep them fully informed. (Tom Adams, Kwinana Refinery Manager) [24]

This incident fuelled community alarm and suspicion and the Alcoa response in the media did little to quell it.

Alcoa advertisement, *Harvey Reporter*, October 2003

Proud to be part of the community.

Alcoa provides strength to the community and touches the lives of many people as a major contributor to the local economy. The Wagerup refinery and Willowdale mine provide employment for more than 850 people. During 40 years in Australia, Alcoa has provided millions of dollars to local community initiatives.

Alcoa is a proud partner of the Waroona Agricultural Show, 2003. [26]

Meanwhile, in Yarloop, community meetings with Alcoa local management had broken down, with Alcoa refusing to provide a 'life of the refinery' guarantee of property protection to long-time residents outside its buffer area.

ABC Radio Interview, February 2003 (excerpt)

John Pizzey, Executive Vice-President of Alcoa: It is no secret we would love to expand in Western Australia. It is a crown jewel of the Alcoa system. It is a major asset for the Alumina Limited .... We have a problem. We created an issue with our community.

> *You can't create issues with communities and then expect to be welcomed back with open arms. I think we are solving problems, working with the government, working with the community to make sure that we meet all their requirements. It is on our agenda to expand. That can only be done when the community is in agreement with that. So we have got some processes still. [27]*

In this senior personnel's comments, Alcoa acknowledges there is a problem and appears here to regard itself as actively working to resolve the issues with relevant parties. At the time the level of community outrage [28] was still largely not accepted as valid by Alcoa. The deep contradiction between the expressed intention to not proceed until they had gained community agreement for the expansion and the community's sense of being afforded no validity by the company was provocative. It was also provocative to link claims of problem-solving efforts to an intention to expand operations at Wagerup, and disingenuous to not realise how this might aggravate the conflict.

The following media report is an example of the order of power Alcoa was exercising [8] at this and all other times in the conflict to serve its own interests.

West Australian, March 2003 (excerpt)

> *A Supreme Court judge has criticised Alcoa for misusing WA's court system just weeks before the alumina company was due to go to trial to resolve a contractual dispute involving a worker's compensation settlement ... Justice Heenan described Alcoa's case as having very little merit and that for the plaintiff to discontinue the action at the last moment bore the appearance of 'taking advantage of the processes of the court' ... Grant Donaldson representing Alcoa said*

the company wanted a discontinuance because it had altered its commercial view of the matter with Mr Thompson ... Mr Thompson, 40, of Success, who made a WorkCover application in 1999, was diagnosed with multiple chemical sensitivity which he claimed was caused by exposure to a range of chemicals at Kwinana from 1998. [29]

Alcoa advertisements, *Harvey Reporter*, 2004

'Our future, your future', signed by Bill Knight, Wagerup Refinery Manager.

Alcoa is working hard to make our alumina refineries even more environmentally advanced. We recently commissioned the CSIRO to review and consolidate all Wagerup emissions data over the last seven years. We asked them to recommend how we can enhance our knowledge to allay remaining community concerns.

The CSIRO affirmed Alcoa has reduced odour emissions at the refinery by more than 75 per cent since 1996. It affirmed that understanding of refinery emissions was unsurpassed globally. It recommended ways forward to further improve our knowledge, which we have started to share with the local community. We are already acting on every recommendation to move forward on air-quality monitoring and modelling in the Wagerup area. Creating a better refinery is good for our future and good for yours. [30]

Meanwhile, the Yarloop and Districts Concerned Residents Group was unsuccessful in gaining Alcoa's agreement that the terms of engagement of CSIRO should include social and community impacts.

Another advertisement at the time claimed [30]:

Alcoa's future is also the future of our employees,

> *our communities and our local businesses. We are currently looking at securing a better future by adding a third production unit at our Wagerup refinery to help meet the growing world alumina demand ... This would mean about 150 new permanent Alcoa jobs and up to 3000 new direct and indirect jobs elsewhere in Western Australia ... Alcoa growing our future.*

Alcoa's contribution to providing employment is a recognised benefit locally. The advertisement perhaps masks the ongoing local grievance raised in many of the community meetings: that very few Yarloop residents had jobs at Alcoa and there was little evidence of direct benefits to the community compared with other nearby towns. Alcoa's Wagerup Sustainability Fund was an attempt to address this criticism. It had mixed results; sentiments in the community were that it was 'too little too late'. The fund, which was established under the trusteeship of the Western Australian Community Foundation in 2004, was hailed by the company and the government as a unique community partnership and social investment, through which local community members could work alongside the Foundation to determine how Alcoa funds would be spent [31].

Sustainability Fund

> *Members of the local community have been working with Alcoa on potential projects aimed at making the Peel and Upper South West regions even better places to live. Two of these projects are the Sustainability Fund and the Enterprise and Learning Centre, both of which have the potential to build sustainable development in the area ... The [Sustainability] Fund will help Alcoa share its business success by linking Wagerup refinery production levels to a community contribution. The Fund would be used to promote*

> *community development projects and capacity-building programs. [32, pp. 3–4]*

Alcoa's Submission to the State Infrastructure Strategy, February 2006 (excerpt)

> *Alcoa is the world's largest aluminium producer and a major Australian exporter. Western Australia is home to the world's largest integrated alumina refining system. Almost 8 million tonnes of alumina is produced by Alcoa in WA each year, accounting for 13 per cent of total world demand ... Alcoa makes a significant contribution to the Australian economy. For every export dollar earned, 80 cents stays in Australia. Alcoa's operations support over 6500 direct jobs and 20,000 indirect jobs, predominantly in regional Australia. Alcoa also contributes over $12 million in community partnerships and sponsorships each year. [33]*

On the basis of this contribution, Alcoa proceeded to outline its claims for improved infrastructure to support its business operations. Accounting for the hidden costs to the public purse of its operations in Western Australia is not part of Alcoa's public discourse [34]. The trade-offs, including those losses occurring in the Yarloop area to the present time, are left out of the picture Alcoa promotes in this crucial document. A benefits analysis without the costs is a biased analysis at its most fundamental level. The state government is party to this power dynamic, and its role is explored in the next chapter.

ALCOA EMPLOYEES

Locally, some of the most outspoken complainants over noise and air pollution and social impacts from Alcoa's refinery were also Alcoa employees – the benefit of well-paid and secure employment for these people has been undercut by losses in their families and community.

***The Australian*, July 2006 (excerpt)**
To the south of Perth, between the Darling Range and the Indian Ocean, lies an industrial operation that is the focus of millions of dollars in spending to create the world's largest alumina plant. And much of the eventual investment will be channelled into producing a social dividend, as well as profits for shareholders and investors ... Alcoa's project involves building a third production unit and upgrading existing plant equipment to improve efficiency and environmental outcomes, delivering an 80 per cent increase in productivity. Capacity at the plant, close to the town of Yarloop north-east of Bunbury, will be lifted from 2.6 million tonnes a year to about 4.7 million tonnes a year – a process that requires the West Australian government to introduce legislation into Parliament to expand Wagerup's production licence.

But the government's decision is more than ordinarily fraught because Wagerup in the 1990s was the focus of a huge public campaign, featuring claims that people in the area suffered health problems. [35]

This newspaper report anticipates government approval for the expansion at Alcoa's Wagerup refinery. The role of the state government in this respect is the focus of the next chapter, but it is worth noting here that while the conflict was still playing out, Alcoa and the government proceeded with the steps required to consider and then approve the expansion.

'Alcoa Charged: Criminal Negligence Case to Answer', *Sunday Times*, December 2008 (excerpt)
Mining giant Alcoa is to face courts, charged with

causing pollution with criminal negligence ... The DEC alleges red dust blew from Alcoa's waste stockpiles which are loaded with highly toxic materials including radioactive thorium and heavy metals ... Alcoa's pollution charge comes on top of a recent health study that found residents living near the Wagerup refinery exhibited higher rates of symptoms associated with chemical exposure than people elsewhere in WA ... But Alcoa and the Department of Health have dismissed the study for which Alcoa paid $60,000, claiming the cancer anomaly was a statistical glitch. [36]

Alcoa and the Department of Health are claiming that scientific research – commissioned by Alcoa – had a technical error. Local people had repeatedly tried to argue that key research undertaken by Alcoa was potentially biased and at times flawed, and had called for independent research. Through their own efforts, independent air sampling by CAPS has provided some of the most significant data of the adverse impacts of 'odour' from the Wagerup refinery, which is at variance with research done by Alcoa.

CORPORATE POWER IN THE GUISE OF SCIENCE

A piece of research published in 2007 by A.M. Donoghue (who holds shares in Alcoa and at the time of the research was an employee of Alcoa in WA) and M.R. Cullen (the same Dr Cullen who came to Yarloop in 2002 recommending Alcoa address the community's health concerns) was used by Alcoa to maintain its dominant story-line. In their article, scientific data support Alcoa's claims of 'no evidence of harm' to community members.

Donoghue & Cullen, Air emissions from Wagerup alumina refinery and community symptoms, *Journal of Occupational and Environmental Medicine* **(abstract)**

Commissioning of a liquor burner at Wagerup alumina refinery gave rise to complaints of malodour and irritation among employees. Subsequently, community members complained about odour and various health issues. Some employees and community members were diagnosed by general practitioners as having multiple chemical sensitivity. After implementation of emission controls, the situation improved; however, community concerns lingered. This paper describes this experience and summarizes several recent investigations including air dispersion modelling, health risk assessment, ambient air quality monitoring, and complaints analyses. It is concluded that refinery emissions currently present negligible risks of acute or chronic health effects including cancer. Communication of these findings has been generally well received, but modifying the perception of risk among some elements of the community has been difficult.

Organizations need to effectively address both technical and perception of risk issues. [37]

The full article is an example of the channelling of company resources into defending its position with scientific evidence and 'other knowledge'. An example of this other knowledge is the argument that the issue was one of perception rather than reality. Complaining residents were typically categorised as 'perceiving' they had a problem as distinct from Alcoa's arguments regarding the scientific 'reality'. Alcoa could use the research to close down any threats to its scientific explanations that did not fit neatly into its finding of 'negligible risks of acute or chronic health effects'.

The authors did concede that, 'in retrospect, with the considerable advantage of hindsight, the initial response to the health concerns of employees and community residents was not as comprehensive as it could have been. This probably contributed to the issue becoming more protracted and discordant than might reasonably have been expected. Sustained adverse media coverage also heightened concerns' [37, p. 1027].

Reporting the outcomes of company-initiated research that used techniques and data from their own sources, the authors allowed for the finding that: 'people attribute incidental symptoms to the detection of refinery odour' [37, p. 1033]. Some attention was then given to negating the credence of these 'incidental symptoms' by surveying other international cases and arguing that communities tend to report health effects when they 'are given a concerning message about an exposure' [37, p. 1033]. They also link this dynamic to a comment, not backed up by evidence, that: 'any adverse media coverage in relationship to refinery emissions is likely to drive this effect' [37, p. 1033]. All other points of contention with the impacted community were then systematically denied or minimised.

The authors intimate and then go on to directly claim that the peaking of health complaints during 2002 and 2003 related to the land management process, with community members being opportunistic (rather than genuinely affected):

Some people attribute nonexistent symptoms to the detection of refinery odour for opportunistic reasons. This is suggested by the lack of a relationship between the complaints of high-frequency complainants and the peak values of NOx. It is also suggested by the sudden substantial increase in complaints when discussions on land management began [37, p. 1033].

The scientific basis to such a claim lacks any robustness. The issue of complaint fatigue apparent to ECU researchers is not considered, nor is the fact that many complainants left the area and that Alcoa's rental agreement with new tenants appeared to disallow complaints.

A lack of scientific rigour is also apparent in the suggestion that community action groups contributed to the problem by lobbying to raise issues of concern.

> *Another important issue is the heightened degree of concern displayed by various community action groups, despite communication of the scientific findings outlined above. This probably reinforces any uncertainty people feel and may result in some people becoming distressed by the mere presence of the refinery [37, p. 1033].*

The authors make their agenda clear in their summary comments.

> *Although we have demonstrated from a technical perspective that there is a negligible risk of health effects from current emissions and the communication of these findings has been generally well received, we have been somewhat less successful in moderating the perception of risk among some elements of the community [37, p. 1033].*

Here the use of 'we' identifies the authors as tacitly aligned with Alcoa's interest in there being seen to be no problems with refinery emissions. Starkly demonstrating this vested interest, the paper concludes with a note about Alcoa's hopes for an expansion. The point is made that this can occur without further odour impacts; in fact, the odours are claimed to *reduce* with increased production capacity.

> *The modelling used emission estimates on the basis of the planned introduction of several additional measures of emission control. The resultant acute*

hazard index, chronic hazard index, and incremental cancer risk contours are similar to those for the existing plant. The odour contours are predicted to contract with the proposed expansion, making the probability of refinery odour detection smaller at any given location [37, p.9].

A manager who worked for Alcoa around the time Donoghue and Cullen published their article describes the folly of trying to hide behind science in these kinds of controversies.

I think for a lot of people it looked like Alcoa didn't care or didn't believe them which is obviously a common thing that can happen when people complain and you present them with science. What they sometimes interpret is: 'we don't believe you'. Anyway it was a big mistake and it changed the whole sort of paradigm down there.

Reviewing the selection of Alcoa's public and private communications in this chapter there is little evidence of Alcoa learning over the course of the conflict. As an Alcoa manager claimed:

If you have a look at the transcript of the Parliamentary Inquiry you'll see Wayne Osborn, the Managing Director, made a comment that Alcoa was slow to respond and he's right and he also sometimes says 'Alcoa dropped the ball' and that's right too. People came to Alcoa and said, 'You're impacting on us, you need to stop it.' What Alcoa tried to do was fix the problem by making the thing run properly and there was technical advice at the time that you just need to get it running sweet ... etc, etc ... a whole variety of things. So Alcoa tried to get the thing to run better. What Wayne says now is what they should have done was just bring emission control in much quicker, straight away.

One of the lessons that the company learnt was that you need to be very careful with changes and the company takes a great deal of care now in the changes it makes because you can have a whole pile of unintended consequences.

The main learning evident is of a company acting on many fronts to grasp the issues sufficiently to be able to defend itself. This defensive action has tended to be achieved at the expense of the credibility of local knowledge, perspectives and experiences. To the extent that Alcoa has not grasped the nature and effects of the power it is exercising to maintain its position of 'truth' and entitlement, the learning needed to inform more equitable relations with refinery neighbours will falter.

CHAPTER 5
SMALL GOVERNMENT AND BIG BUSINESS

There was a feeling shared by community members of being the proverbial 'sacrificial lamb on the altar of progress'. While it was recognised that industry is vital for the generation of employment, income and tax revenue, at the same time it was widely held that the interests and wellbeing of small communities should not be traded off against the interests of profits and big business. In this regard the role of government featured prominently in residents' accounts of the Wagerup controversy. Theirs was a sense of disillusionment with, and betrayal by, their elected representatives who, in their view, protected the interests of industry at the expense of the health and wellbeing of their constituencies.

They experienced great difficulty even in gaining access to government to have their case heard, and were faced with obstacles when challenging the pro-industry stance of Western Australia's political and bureaucratic apparatuses. Their stories reveal suspicions of an unhealthy relationship between the government in its role as regulator and the industry it is meant to control, painting a picture of a de facto protection of industry interests by government.

In this chapter we review community members' perspectives on their elected leaders relating to the Wagerup conflict. We also present the views of civil servants, state politicians, scientists and political commentators who were directly or indirectly involved. First, though, we look at the position of the Western Australian government on economic growth.

DEVELOP OR PERISH

Economic growth and development are the declared goals of many governments around the world; Western Australia in this regard is no exception. Growth is considered the essential means through which wealth and prosperity can be generated and is seen as requisite for long-term sustainability [1; 2; 3]. Proponents of the growth maxim argue that economic expansion will deliver higher living standards and improved quality of life [4], while any associated social and environmental impacts are deemed an unavoidable and temporary cost of growth [5; 6]. Improvements in life expectancy, housing, employment and education levels are frequently cited as evidence for the social efficacy of the growth enterprise [1].

Growth and development in Australia have been a government priority since white settlement [7]. In resource-rich states such as Queensland and Western Australia, development goals have been pursued with particular zeal [8]. At the time of writing, Western Australia is experiencing its biggest mining boom and associated population growth [9; 10; 11].

While growth-orientated policy has remained the darling of Australian governments, the nature of policy making has changed markedly. The growing dominance of economic rationalism across all Australian states and territories over the last 35 years has changed the nature and workings of Australia's political and administrative structures. The newly found appreciation for economic (neo)-liberalism [12; 13; 14; 15; 16; 17; 18; 19; 20; 21; 22; 23; 24; 25; 26; 27] is based on the promise of a freer economy, more openness and less government intervention – in short, small government, efficient markets and big business [28]. This burgeoning economic credo has shaped the nature and dynamics of the Wagerup controversy. Pivotal in its unfolding was the lack of community access to government and political decision-making, and the resistance of governments to playing a more active role during the years of conflict.

Australian state and territory governments have a poor track record in the area of community relations. This can be explained by the marked separation between the country's legislators and their electorates, especially in matters of economic significance [29], a separation based on a negative attitude within bureaucracy and government towards community involvement in the political process. This attitude is a characteristic of the Westminster system and is exacerbated by the new economic perspective. Accordingly, the public is viewed as a threat to the country's growth agenda and to the system of government itself [30], which is built on hierarchical structures and a distrustful political style [31].

Policy making at the state level has been described as particularly 'rigid' and 'hostile to criticism' [32, p. 282], and driven by a fear that public involvement in discussions about the management of the state's resources would lead to the loss of government control and authority [29; 33]. The combination of an aggressive development agenda, a discernibly 'unpublic' political system and an economic approach to government makes for uneasy government–community relations. Conflict arising between the government and members of the community during the Wagerup controversy hardly comes as a surprise.

The Wagerup experience can be understood as an exemplar case of failing to find the necessary balance between the costs and benefits of economic development. In this chapter, we see how the balance was skewed in favour of a narrow economic understanding of the state's development needs at the expense of those who are meant to be the beneficiaries of economic advancement. The widely recognised shortcomings [34; 35; 36; 37; 38;

39; 40] of the reductionist, economically rational lens of government are writ large when applied to the complex and messy socioeconomic and socioecological problems at the core of the Wagerup conflict. The separation between the state and its people is also manifest in the trust issues that emerge in the community accounts.

THE UNFULFILLED PROMISE OF REPRESENTATIVE GOVERNMENT

In our conversations with community members it became plain that residents had a clear understanding of their expectations of the state government. Indeed, the community's formula was quite simple. Residents saw the key responsibilities of government as the protection of the interests of the state and its people.

> ... they're supposed to be watchdogs. (Yarloop resident)

> [T]o see that our society stays afloat, [the] economy should be looked after ... Development needs to be sustainable. (Yarloop resident)

> Keeping an eye on [industry] ... to put the rules and regulations on them; ... it's got to be safe for people to live in; if they want to mine this place they also have got to make it safe. (Yarloop resident)

> ... certainly, the rights of their taxpayers and their residents need to be protected. The environment needs to be protected. (Yarloop resident)

These statements make explicit community members' acceptance of the need for development and for it to be a government priority. Accordingly, elected leaders ought to ensure long-term economic growth and the development of the state. This, however, is not to be seen as an unqualified mandate for growth at all cost. It is equally important to keep development in check, ensuring that social and environmental interests are not compromised.

The expectation of 'being protected' from the impacts of development featured prominently in residents' discussions about the role of government.

> *I'm not naive to think that big business doesn't have to progress; there's a lot of people who live in Yarloop who work at Alcoa and that's necessary, but still at the same time there has to be protection. (Yarloop resident)*

It is this level of protection, however, that residents felt was not afforded to them by government, leading to the disappointment and disillusionment of many with their political leaders and the political process as a whole.

> *I've lost a lot of faith in the government's protection. I've got a lot of reasons to distrust the government. (Yarloop resident)*

> *Well, I have to admit I have lost a great deal of faith in the political processes of this country. (Yarloop resident)*

> *I was inclined to believe my government and no longer. I can't honestly believe that our government is going out of its way to protect its citizens. (Yarloop resident)*

One of the key criticisms locals levelled at government was the ignorance displayed towards the plight of people living around Wagerup when health problems first arose. According to residents, the various government departments displayed a lack of interest. Nobody seemed prepared to listen.

> *The Government's not going to listen to you. (Yarloop resident)*

> *I emailed ... a couple of the ministers ... no response, nothing, not even the courtesy of a form letter, nothing. It's total dismissiveness. (Yarloop resident)*
>
> *I got no replies ... bugger all actually...maybe one or two out of eleven. Same old story! There's only so much you can do. (Yarloop resident)*
>
> *I would have expected at least somebody to show up at my front door and want to talk about the issues that we are going through. (Yarloop resident)*
>
> *I just poured my heart out to this man. I said: 'I don't know if you're aware of what's going on here,' and never heard a thing back from him. (Yarloop resident)*
>
> *If it's just apathy well then that's unforgivable as far as I'm concerned; if it's just apathy from these departments then I find that horrendous. I'm not sure what it is. But there was definitely a reluctance in the beginning. It was almost cynical ... I don't know what they thought we were but they certainly weren't very helpful ... (Yarloop resident)*

In fact, the government and its departments appeared reluctant to become involved at all.

> *The government refused to become involved. (Yarloop resident)*
>
> *[T]he worst part about it: everyone knows what's going on but no one will do anything. (Yarloop resident)*
>
> *It seems to me that when you go to a government agency with a problem this white wall goes up. Then you have to prove to them that there is a problem before they will react and I think they could be a lot more proactive. (Yarloop resident)*
>
> *Now they stood back when they should have done*

something. They were informed from many different people and groups at that time, and yet they did nothing. (Yarloop resident)

All these reports, all these things that have happened, the amount of information that goes back to the government, and yet there's silence. Nothing happens. So where is the government? Good question. (Yarloop resident)

Disappointment with government was not party-specific, as both major parties were found equally wanting.

The government, they came down and they paid lip-service; the Liberal government of course would never have done anything because Charlie Court put it there ... So the Liberals would never have done anything. (Yarloop resident)

... then [the Liberals] passed it onto Labor, thinking that they would have to deal with it. Well, they were actually more adept at passing the buck than the Liberals were from where I'm sitting because ... nothing has been addressed at the end of the day. (Yarloop resident)

There was much speculation about what lay beneath the reluctance of successive governments to become involved. Many residents saw a political agenda at work. In their view the Wagerup controversy spelled political or professional suicide for anyone getting too close to it. Wagerup was considered simply too hot.

They're too scared, they know but they don't want to do anything about it because they don't want to change jobs ... We don't want to find out because we don't want to know because then we will have to deal with it. (Yarloop resident)

> ... when there's clear breaches of the regulations ... then prosecutions but that's a dead-end career path for whoever pushes to the prosecution at present. (Yarloop resident)

> It's just too hot to touch. Nobody wants to know. (Yarloop resident)

In the end, people felt the government left them with no choice but to take matters into their own hands, almost forcing them to mobilise.

> The government has forced [the community] to form a group that gives them a voice that they can be heard and really you've got the community having to spend their own resources, their own time and their own effort and everything in there when they shouldn't have to. (independent consultant)

The political sensitivities surrounding the Wagerup refinery were seen by residents to be reflected in the rapid turnover of departmental staff.

> [A]s soon as people in government that are there for the right reasons start to understand and they try to do something they get shifted and they get either sacked or moved out. (Yarloop resident)

> ... every time a member of the Department of the Environment and Conservation, every time a member of that group brought up a contentious issue, he was shuffled sideways or sacked. I find it rather coincidental. (independent consultant)

The constant changes meant that community members often had to start back at 'square one' in their dealings with government departments. There was no consistency and no apparent accountability.

> *... if you've got a complaint it takes you days before you can even talk to somebody. Nobody knows anything. Oh, it's this department, no it's that department, no it's this department; no, we don't have anything to do with it anymore we can't help you, because it keeps changing. (Yarloop resident)*

> *I mean even the appeals convener, he's not there. He's got nothing to do with that department anymore, has he? I mean people open up and they just walk away. It's just too hard ... We can't deal with it. We don't know where to start. They wouldn't even know where to start. Where would you start? Where would you bloody well start? (Yarloop resident)*

> *I just got this email back to say 'basically just fucking leave me alone, I've got nothing to do with this anymore', and that was it. No explanation, no nothing. (Yarloop resident)*

Ministerial portfolios also changed regularly. Under the Carpenter Labor government between 2006 and 2008 the environmental portfolio saw four changes of minister.

> *... we had the Minister down here with a meeting. We all sat and they told us right at the beginning he's not answering questions, he just wants to hear people's stories and I must admit that was one of the quietest meetings that I've ever been to, and we all got up and told him what was happening, how we felt, and what happened to him? He was promoted. He seemed really to be listening and be on our side and then a couple of months later I kept saying to [name omitted to protect privacy], 'What's happened, we haven't heard anything,' and he said, 'oh, he's got another job, he's been moved' and it's things like this that make*

> us think, well why, what's behind it all? It just makes you suspicious all the time. I've had it up to here with meetings, I've stopped going. (Yarloop resident)

These frequent ministerial changes during a time of far-reaching government decision-making resulted in a blurring of the lines of accountability, exacerbating the sense that residents were left without recourse. Community members started to question the nature of the ministerial reshuffles for they seemed to follow pertinent ministerial decisions. Environment Minister Mark McGowan's approval of the refinery extension in 2007 was soon followed by his transfer to the education portfolio, with Anthony McRae taking on the Ministry for the Environment. The joint departure of former Premier Dr Geoff Gallop and Environment Minister Dr Judith Edwards at the height of the conflict were also considered suspect.

> If you will pardon my paranoia, I find it inconceivable that the Minister for the Environment can quit the same week as the Premier for health reasons without some background interplay ... (Yarloop resident)

WHO IS REGULATING WHOM?

Beyond questions of accountability there were also grave concerns about the adequacy and sufficiency of the regulations governing what was considered a high-impact industry.

> You'd have to say that how they are managing and regulating is totally inadequate and inappropriate. (Yarloop resident)

By Alcoa's own admission, environmental regulation in WA seemed more relaxed than in other states.

> ... [in] Western Australia both from a mining and a refinery perspective I don't think there's ever historically been a view that there are any real significant health risks within those businesses ... (Alcoa manager)

> [T]he licence to operate was less constraining in Western Australia. (Alcoa manager)

> Western Australia control regimes are not as tight as they are in Victoria for example. (Alcoa manager)

From a community point of view, regulation was inadequate and enforcement too weak. People doubted the ability, people-power and expertise of government departments to effectively monitor the industry.

> There's nobody in the government or on the EPA ... There is no one. That's why it takes so long for them to process and find and do all this sort of shit. There's no one there that can disprove them or can even ask them the right questions ... They don't actually know anything about the actual scientific stuff. (Yarloop resident)

> ... there seems to me to be this 'We don't have time, we don't have enough resources'. There are always excuses why monitoring isn't done or whatever but if that's the case, if that's the only reason why they're not doing their job, then industry should be being levied to make sure that the environment and the people and the animals in the environment are protected. (Yarloop resident)

> [The Environment Department] have got no idea about industrial regulations. They might be good on forests and trees, biodiversity and all the rest of it, but when it comes to regulating industry, what the hell do

they know? (Yarloop resident)

A particular bone of contention throughout the Wagerup conflict was industry self-regulation, a dominant feature in the policy 'toolbox' of neo-liberal governments. Following the principles of market-based environmentalism, industry self-regulation enables companies to monitor and report on their own environmental management performance, while the role of government is reduced to that of a rule-setter and umpire. The guiding principle underlying this approach is a strong faith in the workings of the markets and the ability of industry to become part of environmental solutions [41; 42]. Its attraction lies in the flexibility given to industry to achieve environmental goals and the cost advantages to government when compared to conventional command-and-control approaches in environmental regulation, which require considerable administrative input in the form of monitoring and enforcement. The self-monitoring approach is also premised on the tenet that economic growth and environmental protection are reconcilable [43]. Unsurprisingly, this approach is not without its critics [44].

From a community perspective, industry self-regulation amounted to Alcoa being left to to be judge and jury on its own performance.

> *It's like letting the cat look after the mouse. (Yarloop resident)*

> *Like if you had your own business and you are chucking all these nasties out there, you are not going to say in your reports that I chucked out so much rubbish. What is going to happen to your business? They are going to close you up. So you are just going to go tick, tick, tick, everything is fine. (Yarloop resident)*

> *Self-monitoring is not a good scenario. We feel that all along they haven't been supervised well enough.*

I think their self-monitoring is not reliable. (Yarloop resident)

They've got their opinions about whether it should be there but let's put it this way: the first testing they did they were doing them on their own. Well, what would you do? Would you judge against yourself? (independent consultant)

The point that's always stood out with me is that regardless of what's being done, Alcoa seem to be monitoring themselves. (Yarloop resident)

I have absolutely no faith whatever in self-regulation. (Yarloop resident)

These comments show that industry self-regulation enjoyed little support among community members, who saw in it a 'carte blanche' approach for industry by government and as evidence of 'regulatory capture' [45], which allegedly opened the floodgates for corruption and company misdemeanours.

... the whole lot of corruption to do with industry in WA can be brought down to one simple thing and that is self-monitoring. When industry were allowed to monitor their own situation that was rife for corruption. That is just made for it. If you've got the results of two samples and one is good and one is bad, which one is going to be sent in? It's [...] fairly obvious ... (Yarloop resident)

The only way they can fix the environmental problems in WA is to remove self-monitoring ... That's the only way they can clean it up. Until then we're going to have corruption but you need a very gutsy politician to do that. (Yarloop resident)

Residents also saw an injustice in self-regulation. In

their view, Alcoa's size and econo-political influence allowed the company to be self-monitoring, a courtesy that government did not always extend to smaller businesses.

> *So where's the justice in all that? The small people have to fend for themselves, fight for their own rights with the government who are elected there by the people, to look after their rights and to stop people like this trampling on the small people, creating health problems, environmental problems, water table problems, it's all part and parcel. (Yarloop resident)*

> *The government should be there regulating these people. Like if there's a small [business], like the Hazelmere abattoirs, because they are a small group, the government jumped on them. (Yarloop resident)*

Of particular concern were questions of power and control relating to the relationship between Alcoa and the government.

> *We've had employees of the Department of Environment refer to their Department as the Department of Alcoa Protection. (Yarloop resident)*

> *It seems that whatever Alcoa says the government has to do, they're too scared to disobey. (Yarloop resident)*

> *I think Alcoa's got all the control. They tell the government what to do.*

Reminiscent of Australia's climate change policy under the former federal Howard government [46; 47], it was widely held by Yarloop residents that Alcoa was ultimately in charge of decisions and events during the years of conflict and largely able to determine the manner in which the company was regulated and its operations were governed.

[I]t's surprising that the level of acceptance by the regulators and the government, given the impact that this profit-making enterprise has had on the local and broader community for that matter.

It's not an independent source investigating ... they're investigating in conjunction with Alcoa, and Alcoa are in a position to just say those facts, and people are accepting those figures from Alcoa ... emissions and things like that.

The government and Alcoa put this together when you read the previous history of what was going on. It was a plan that was developed by Alcoa and the government on how to proceed, their way forward, of dealing with the community issues as a result of the refinery impacts, so it was staged and planned.

... they will only do what they want to do unless they're forced and that's been made evident so many times now, and the problem being is the government's weak-as.

Alcoa's controversial Land Management Plan was a much-cited example of the government's hands-off approach to the Wagerup conflict. The government's refusal to install a formal buffer zone around the refinery enabled Alcoa to introduce a 'voluntary' land management regime. Because it was voluntary the government refrained from becoming involved when residents fell foul of the terms of voluntary relocation as proposed by Alcoa.

Well, the effect now is going to be greatly magnified because of how they have chosen to allow Alcoa to deal with this and that shift in population was due to Alcoa's voluntary buffer acquisition, and the word voluntary is the reason the government haven't got involved. They've claimed that it's a voluntary

> *undertaking by Alcoa so therefore they cannot become involved. (former Wagerup resident)*
>
> *The government refused to become involved even though ... Alcoa's reports have actually now shown that properties were being undervalued by up to 40 per cent. There were two reports, both funded by Alcoa which found a similar level of undervaluing ... So, despite people going to the government and saying 'Hey listen, we're being ripped off here but we haven't got a choice, we have to sell to Alcoa, we need your help,' ... they're saying 'Well, we can't because it's voluntary.' (former Wagerup resident)*

The government was seen as being weak in its approach to regulating Alcoa and regarded as too close to the company.

> *I'd stand in front of the government and tell them that too: you're all sleeping in the same bed with Alcoa. I have nothing to hide. If they want to put me in gaol they can put me in gaol. (former Yarloop resident)*

Local anecdotes abounded of alleged irregularities and collusion between the government, government departments and Alcoa, leaving a sense of a community doing battle with both its corporate neighbour as well as the government.

> *You only have to look at ... those people that did their homework on Alcoa for the government ... those that were sent down to check on Alcoa and find out what their problems were and report back to government bodies – where do they now work? ... for Alcoa ... Conflict of interest? I think so. (Yarloop resident)*
>
> *... we need to have an independent inquiry that looks at firstly industry connections, because WA is a closed*

sort of a shop. There's a revolving door there and there has to be some monitoring of that ... I believe that there has been some form of inappropriateness within the department due to the industry connections. (Yarloop resident)

... there were two factions in the Health Department and some said 'yes we can see that there is an issue' and some were in total denial, but it's really interesting that a couple of those people that were in total denial are now working in the mining industry so that's how that goes sometimes. (Yarloop resident)

In Chapter 2 we described the active support by the government of the time for the establishment of the alumina industry in WA in the face of considerable community opposition. Over the years, the level of political support for the industry has not wavered, attesting to the continued development focus of successive state governments and their efforts to support growth in employment and investment. The quote below by WA senator Kate Doust illustrates the government's support of Alcoa and its belief that local problems have been solved [48].

While some people may not be happy with that situation having occurred – some members of the community do not want the expansion to occur at all – there are other people who live and work for the company who are looking forward to the expansion. We need to keep in mind that the company is a substantial employer in our state. It is a company that has put into the local community huge amounts of dollars to support activities in the community. As I understand it, it also has a long-term sustainability program in the area for the next 20 years. There are lots of pluses associated with this. The problems people have referred to were dealt with during the 2004 inquiry by the Standing Committee on Environment and Public Affairs.

Doust's statement reflects a strong faith in the drivers of economic development, the benefits commonly assumed to follow industry expansion. But while claims to community benefits and problem resolution are asserted, they are not established [49]. Alcoa's much publicised philanthropic spending and community investment [50; 51] were met with scepticism among community members, especially since spending amounts were tied to the company's ability to expand. In other words, community funding was contingent on Alcoa realising production increases and, as far the community was concerned, increases in pollution.

[W]hen you ... meet these champions of community development ... you realise that it's all a big farce. (Yarloop resident)

At the same time, residents were regularly being told by company management, and by government and its departments, to 'let go'; that the conflict was over and that the community ought to move on.

That's in the past, let's move on. That's the only response you get. (Yarloop resident)

In fact, to many residents it seemed that they have gone full circle, ending up back where they were more than ten years before.

[As far as] I'm concerned, th[e health] issue has been put to rest through the Parliamentary Inquiry. Alcoa has admitted that it has affected the health of the community, or some community members, so that's

the Parliamentary Inquiry and also to the Wagerup Medical Practitioners Forum. They've admitted it in the newspapers, so really I'm frustrated because we keep going this full circle with the health issues. Where it's gone from being recognised to now we're claiming again, and the biggest battle in all of this has been to get that acceptance of the health issues because that is the main reason this issue exists in the first place. (former Wagerup resident)

Residents certainly agreed that inquiries had taken place and that volumes of information had been collected and discussed in connection with the Wagerup refinery. But they would not readily agree that the issues on the ground had changed; many believed that key findings from various processes and inquiries were ignored by government who, in their view, failed to act upon the information.

There were recommendations made in that inquiry that have never been followed up, so there's apathy there still. (Yarloop resident)

... the government bureaucrats might say that it's been addressed and certainly Alcoa's corporate management may consider it addressed but at the end of the day when the ones who are being affected don't consider it to be addressed ... How can ... [the government] accept it that easily, which is what they've done? They just continue to ignore those [...] affected. (Yarloop resident)

I know the Medical Practitioners' Forum has made recommendations to government which have been ignored. (Yarloop resident)

Qualified doctors who are essentially smeared by the company and their advice ignored, and then when a bad decision is made the people who make the decision turn around and say, 'Well, we didn't have any advice.' Bollocks, the advice is being screamed at you from the rooftops, you're just simply choosing to be deaf. (Yarloop resident)

... the government and everybody let this happen to these people. The others pissed off, and the ones that are left are the ones that are broken, and nobody gives a shit. (Yarloop resident)

Though the conflict was ongoing in the eyes of residents, the government considered the case closed, and chastised community members who continued active campaigning 'to receive justice'. Community agitation was considered unwelcome, even dangerous, evident in a letter from former Premier Alan Carpenter in response to ongoing community complaints [52].

The government believes the undertakings given to the state by Alcoa are fair and equitable and address the concerns about the refinery operations. Such undertakings by a company are unprecedented in Western Australia's industrial experience. There are limits, however, to the reasonable response that can be expected, even from a world-scale commercial operation, before the impact becomes a serious disincentive to further investment and is detrimental to the economic and social benefits that the local and state community as a whole will obtain.

All in all, as far as residents were concerned, the message from government was loud and clear that their government was 'putting ... revenue ahead of the community' (Yarloop resident).

This government is saying well, there's only so much we can impose on this industry because we want their money. (independent consultant)

TAXPAYERS VERSUS TAX REVENUE: WHOSE GREATER GOOD?

Over the years, residents came to realise that little support could be expected from the government. Attempting to make sense of this, many residents concluded that the government was 'more concerned about expanding the refinery and getting more jobs and more revenue' – protecting the interests of 'the corporate body rather than the local community' (Yarloop resident).

A key explanation for many as to why the government chose to side with industry and not with its people, was that it was blinded by the tax revenue.

It all boils down to money. (Yarloop resident)

I don't know how you fix it when you've got a government so greedy and forgetting [...] they're put here [...] to look after the people who put them there. (former Yarloop resident)

The government chooses to ignore it. All the groups they've put together to have a look at this; the Health Forum, Doctors' Forum, the passing of this third unit ... [they] chose to ignore all their advice and went ahead and gave it to them. Why? $11 million more for the government! (Yarloop resident)

Residents saw in the actions of government a calculated manoeuvre, a political cost–benefit analysis determining the economic pros and cons of an expanded refinery. Local social aspects did not feature prominently in these contemplations.

It's the same [with the] Wagerup [extension], for health reasons alone it shouldn't get the tick of approval but the government has because, 'hey, there's going to be more jobs, more taxes', and that's their number one

> *up the top. Same as Alcoa: profit. The government's looking at the same thing. (Yarloop resident)*

> *They weigh up human environmental cost against the positive outcomes in profit and I guess the money side of it for the government and the company comes out on top. (Yarloop resident)*

> *I think mainly because [...] there's a big economic benefit to the state [...] they allowed things to just proceed as normal ... and I think we were the guinea pigs that were ... not listened to. (Yarloop resident)*

Many residents were eager to explain that they were not anti-corporate or anti-development. They sought accountability and were not interested in shutting down Alcoa. They were disillusioned, however, with their experiences with the political process and the lack of support they received from the government and its departments. While there was faith in the possibility of arriving at a political solution to the Wagerup conflict, residents were pessimistic about the likelihood of it occurring. Reference was consistently made to the need for a strong government. However, based on what they saw as unhealthy econo-political entanglements, there seemed little prospect of the government freeing itself from corporate influence.

> *That's our concern ... Unless you've got people in power who are able to take the bull by the horns and actually act on it, it's just a waste of time. (Yarloop resident)*

> *I understand the government ('we can't afford to tell Alcoa to piss off because it's revenue.') The country has to run. It's like here, we've got a household, we've got to make money to survive, and a government is exactly the same but aren't people's lives worth*

> *anything or are we only just a handful of people not worth worrying about? (Yarloop resident)*
>
> *[Alcoa] are getting the backup from the government because they're employing a lot of people ... That's the way we feel. We feel the government's not going to do anything. I think the government is on their side. (Yarloop resident)*

Trust in government and politicians in Australia has traditionally been low [53] based on political apathy and a sceptical, even cynical stance towards the country's institutions and political leaders [54]. Trust in government encompasses a people's confidence in government and its institutions, policy making and political leaders. It can broadly be understood in terms of people's support of government spending, faith in law enforcement, perceptions of honesty, and faith in democratic processes and institutions [55; 56].

> *It's not a place where you can trust the government. (Yarloop resident)*
>
> *You can't believe the government because the government doesn't care. All they're saying is, 'Yes, we're going to let Alcoa do this.' (Yarloop resident)*
>
> *It's just a natural distrust for government agencies and I thought I had moved past that; I thought that the government was going to protect us and I was perhaps a little naive in that fact [as] we are not being [protected] ... I find that a little bit of a slap in the face because we are productive people. (Yarloop resident)*
>
> *... the government has never, ever helped the community. They were silent all the way through this. (Yarloop resident)*
>
> *It's hard to have a lot of trust [in government],*

> *because, historically, there's no reason to. (Yarloop resident)*

The notion of 'sacrifice' was another consistent theme. Overwhelmingly, community members spoke of feeling 'exploited' and traded by government for the greater interests of the state.

> *Australians are supposed to be the greatest asset of Australia. The people that live here are supposed to be our greatest asset and yet apparently aluminium is. (Yarloop resident)*

> *I mean, progress changes everything, our progress is fine but we should still be able to live and exist [in] a fairly free lifestyle down here. We are after all in a beautiful piece of the countryside [...] I kind of feel like all the time [...] we are just being bombarded from every direction. It's like you have to be on guard all the time because we are under attack. (Yarloop resident)*

> *Yarloop was just a little poor old town with a few hundred people there and Alcoa is this great big mining company and billions of dollars, and money speaks all languages; the government and Alcoa are all sleeping in the same bed, they work together; you haven't got a hope in hell. (former Yarloop resident)*

> *Yarloop was there long before Alcoa. Alcoa just came in like a big bully and took over and who cares about these people, who cares if this person has got an unusual disease and has died, who cares if this one here has got breathing difficulties, who cares if this one ... number one straight down the line, money, money speaks all languages. (former Yarloop resident)*

The demise of Yarloop was seen to have become the inevitable cost of progress, and residents saw themselves reduced to a 'disposable' community [57]. They had to come to terms with the fact that their government was not going to protect them, that its dependence on mining and refining seemed to override the concerns of a relatively remote and largely invisible community. For the residents, progress for the state did not translate into progress for all. They knew it came with weighty social and environmental price tags for those living too close to the wheels of progress. While most members of the community accepted the need for some trade-offs to be made in the context of the industrialisation of their region, there was an expectation that these trade-offs would be openly acknowledged and that the community would be protected from undue harm and compensated adequately for any losses incurred.

VIEWS FROM PRIVILEGED OUTSIDERS

In the remainder of this chapter we invite the voices of informed outsiders including politicians, scientists and public servants who were close observers of, or active participants in, the Wagerup conflict. From their privileged vantage points they provide further insights into the conflict dynamics. Their views are organised around three interrelated elements that relate to the role of government as legislator and industrial watchdog: the adequacy of legislation; the quality of independent scientific expertise; and the handling of dissent. As in the rest of the book, names of the speakers are not given due to the ongoing status of the case. Note too that while ministerial views were repeatedly sought, invitations were consistently declined.

1. The adequacy of industry regulation

A key observation made by bureaucrats, scientists and politicians alike was in regard to the inadequacy of the legislation governing heavy industrial facilities in the state. There was widespread agreement that the role of government is to put in place rules that govern industrial activity, since the commercial self-interest of companies was seen to render social and environmental best practice unlikely. In other words, a strong government and sound regulation were seen as essential.

> *Industry will only go to the extent that they legally have to but the community will go beyond that because they are obligated as humans [...] to care for the environment ... Industry will never cross that line. They will only do whatever is needed of them and until they're forced to do more and that can only come from government, they're not going to do it. (government department staff)*

> *Corporate codes of practice are not enough to ensure either environmental protection or occupational health and safety. (state parliamentarian)*

> *... I think that ultimately the responsibility rests with the government to have better provisions, better legislation ... (independent consultant)*

> *Self-regulation is bloody airy-fairy bullshit. You have to have clear processes, clear regulations ... You have to have reason within that but you do have to have a regulatory body, that's government, and they have to be able to ensure compliance and if there isn't compliance, deliver enforcement and if they don't do that they're wasting their time. (government department staff)*

In the case of Wagerup, industry regulation was considered weak and flawed, serving the interests of industry and government but failing to protect the interests of communities.

... it's true to say that through the State Agreement Act we created laws which actually served the whole interest of the company and the state ... and this was Sir Charles Court, following that Richard Court and then every state Premier, including Geoff Gallop and Alan Carpenter, signed off on expanding the state's industrial activities through these Agreement Acts. (state parliamentarian)

State agreements are essentially contracts between the government and industry that are ratified by parliament. Since the 1950s, state agreements have been used extensively to foster resource developments in the state. Such agreements are attractive to industry because they provide the long-term certainty of resource access and land tenure vital for corporate long-term planning [58]. State agreements also serve the purpose of advancing the state's development agenda. As noted by two state politicians:

... historically what was happening was the states were competing amongst themselves to get large projects because that's the way they saw that you were generating economic activity and jobs and so on and these State Agreement Acts just basically trumped everybody. So ... there was a kind of social contract that says we want the financial and economic benefits and that we're prepared to sign off on the arrangements and we are going to accept a whole lot of risk. (state parliamentarian)

The source of this problem goes back to the 1970s when Western Australia was governed by a Liberal government, the Court Government, which had a philosophy that all development was good; 'development at any cost' we used to call it back in those days. (state parliamentarian)

The nature and purpose of these acts, however, were criticised because they seemed to place industry outside the legal framework that commonly applies. As a WA senator explained:

> *It's easy to operate within the law if you construct a law around doing something which is fundamentally unjust or not sensible or not in the public interest and we systematically perpetuated that by these State Agreement Acts, which should all be outlawed, because in effect what they do is they put the agreement and this arrangement, this commercial arrangement, this contract, outside of the normal regulatory arrangements of the state. They said, 'we've got all the regulatory arrangements and then we've got this agreement' ... where else does this happen in the world that the state forms a formal agreement, a statutory parliamentary agreement with royal assent to go and do something; why wouldn't the company just operate in the normal laws of the land? The question of whether it's legally within the letter of the law becomes blurred by this unusual relationship that was created by the State Agreement Act and the collusion between, if you like, Liberal and Labor [governments] and unions and business to promote those developments in spite of the public interests, or consequences.*

Moreover, the security granted to industry under state agreements was considered to be reflected in the state's pollution laws, which govern industrial activity. Echoing community perceptions of government regulation, regulation in the case of Alcoa was widely felt to be ineffective and inadequate, in that it failed to match the complexity and sophistication of the industry it was meant to govern, and because it gave Alcoa too much control over its environmental management obligations.

> *I think also another key issue ... was the absolute inadequacy of air-quality guidelines. I mean, go back to the 1980s, they don't have specific air emission guidelines for volatile organic chemicals, let alone the incredibly complex issue of multiple chemicals, and this again shows the inadequacy at this stage of scientific knowledge about impacts on the body of being exposed to literally hundreds of chemicals that are known to have a toxic effect on the body ... No one actually knows and there are no air emission guidelines which try and deal with that matter and yet we know that there are at least 361 different toxic chemicals in the emissions at Wagerup. (government department staff)*

> *Government should be the one saying to industry this is how we want you to regulate yourself, we'll be performing our own checks, rather than your own people doing the test themselves, so you've got an industry here which is effectively self-regulating. (government department staff)*

Overall, the regulatory frameworks governing Alcoa's operations were considered a 'light touch' and ineffective in terms of protecting communities from industry impacts. As expressed by one scientist:

> *I see it as a failure of our institution in the legislation and process of the environmental protection that's meant to be in place.*

> *There's zero tolerance in some areas of law but not zero tolerance when it comes to environmental conditions.*

The government's stance on industry regulation was considered to be consistent with its very overt pro-development bias. Unsurprisingly, community concerns about industrial activity and associated health impacts were seen to threaten the state's growth agenda and were considered a political inconvenience. It seemed more than likely the community would face political resistance when trying to have its case heard by the regulator.

> *When the liquor burner was commissioned in 1996 ... it is very clear that ... the Department of Mining ... of course being the regulator in the first place of the air emissions from the refinery ... was extremely reluctant to take the issue seriously because it was very inconvenient to everyone – because everyone preferred a business as usual scenario where a major company could go on about its business of its mining plans and so on ... the air problems have been very inconvenient for government because otherwise it would be a happy story of economic development and dollars rolling in through royalties and blah, blah, blah. (state parliamentarian)*

> *I don't think that government or Alcoa are interested in resolving the real concerns of the community. I think they're primarily interested in keeping the business going and keeping the revenue flow going. (state parliamentarian)*

Public servants in particular spoke of a reluctance by government to engage with the community and of a lack of willingness at the departmental level to investigate the community claims.

> *It's been very tempting for everyone to not believe that the air pollution problems are real and that of course has been compounded by the complexity of the*

science around air modelling and the science around actually capturing these specific pollution events, which are ... the cause of multiple chemical sensitivity and other specific poor-health triggers. (government department staff)

It is hard to find anything and also one doesn't want to find anything. I'm not saying that people deliberately set out not to find anything because there are scientific and technical issues that made that difficult. (government department staff)

There were ideological issues that also prevented people from, at certain times and places along the many years that this story has evolved, where there hasn't really been enough commitment to finding the problem ... (government department staff)

Not only were government departments found to have been reluctant to give credence to community complaints, they were also seen to be technically inadequate to deal with the complexity of the science of alumina refining, in part – as one political commentator notes – due to poor resourcing and low staff numbers.

... you've just got the Department of Environment and Conservation which has been significantly resource-strapped for years, in terms of both money and personnel, and there's such a myriad of laws and regulations. There's a document that's come out; it's an internal document that I've seen a draft of where they've basically acknowledged that they cannot, at the moment with their current level of resourcing, enforce licences, monitor licences, monitor compliance with ministerial conditions because of poor resourcing. (government department staff)

This view was also shared among parliamentarians, departmental staff and members of the scientific community.

> ... the governments of the day have consistently not resourced – the Department of Environment in particular – to have adequate monitoring, to have the staff on the ground and go out and actually do any kind of independent monitoring. (state parliamentarian)

> ... but also because of a lack of people with the necessary expertise. Experts in the field expressed grave concerns about the scientific skill shortage (independent consultant).

> Our government didn't have the right departmental structure and academic depth. (government department staff)

> ... the department was ill-equipped and lacked the available internal experience required to better manage this industry and regulate this industry, rather than the industry self-regulating itself and providing reports to our government departments saying ... 'we've done the test and everything seems to be working well' ... I think the government needs to have the right set of skills in place in order to set the agenda for these industries to pollute to an acceptable environmental level. (government department staff)

> It's a real failure of this government, not understanding the full implications of the process by-products [of] this industry [are] having on our environment, let alone what [they are] having on our community. Even now I don't ... believe that our government has a full grasp of the potential health impacts [on] the constituents within the emissions ... or ... on our community. (government department staff)

In other words, the departments in charge of regulating and monitoring the industry were considered incapable of testing the veracity of the company's emissions data. Despite the need for a better scientific understanding of the pollution issues that have fuelled the Wagerup conflict for many years, opportunities to get independent third-party advice were rejected. Despite strong support for an independent air-monitoring study by CSIRO, the need for which was also recommended by the Parliamentary Inquiry [59], neither Alcoa nor government supported such an investigation.

... [there] should [have been] an independent study of the air monitoring by CSIRO. The government rejected that recommendation. Now as far as I'm concerned if there's nothing to hide and there's nothing wrong, then there's also nothing wrong with having an independent third party doing that work with their own equipment, choosing their own monitoring sites, choosing their own methodology. The fact that the government wouldn't agree to that and obviously the company wasn't keen either, indicates that they're not willing to have a recognised authority in these sort of areas take an independent look at it. (government department staff)

The Department of Environmental Protection did not have the expertise to really crack the problem. However, they could have obtained that expertise, and I think that they failed to give the problem sufficient credence to bother to spend the money ... (government department staff)

Overall, the government and its departments were largely seen as both unwilling and unable to engage in earnest with the pollution and health complaints at

Wagerup, resulting in the 'government fail[ing] to take control of the situation,' as a former state politician put it. Unsurprisingly, the community 'didn't have any confidence in the government system because the government wasn't as well versed on the implications this industry was having on the community,' as one scientist suggested. In addition, government and its departments were seen to be complicit in the demise of Yarloop in light of the fact that 'the trade-off ... between the risk to the health of the individuals and the benefits to the rest of the Western Australian community appear[ed] to be ... accepted'. It was, as an independent consultant said, a case of another 'Wittenoom, a town that basically was encouraged to die.'

2. The availability of independent scientific expertise

The things that come to mind are control; [Alcoa] controlling community consultation, controlling their image, controlling bad news when it occurs, spillages, leaks, you name it, controlling outcomes from government and controlling the media. (political commentator)

Two aspects of note regarding departmental science are the relative void of independent scientific expertise in Western Australia in the area of industrial pollution, and what was regarded as a corporate strategy to 'white-ant' the limited expert realm. Government departments in Western Australia face great difficulties when seeking to recruit technical experts, especially in highly specialised fields. The state's recent resources boom highlighted the short supply of technical experts and the strong competition between the private and public sector in the recruitment of talented staff [60; 61]. Alcoa was seen to strategically target scientific expertise within government departments, universities and research institutions by way of commissioning research, funding academic positions or recruiting experts, leaving WA devoid of independent, non-Alcoan science. As one former state politician explains:

> *One of the really sinister aspects of the way that Alcoa operates is that you see them moving into areas where there may be some independent scientific expertise and behaving in ways that would ... seem to be ways of spending a lot of money to court influence in those areas, by setting up university departments, new chairs in university departments, all the while wanting to buy good influence ... it's actually fairly hard to find anyone with the requisite knowledge and authority to challenge Alcoa, so there was that level of difficulty for the environmental regulators to deal with, too.*

This corporate funding of academic research and positions as a form of monopoly-building, was seen to serve the purpose of quashing opposition and minimising dissent.

> *One example that comes to mind [...] was at Curtin University. Alcoa, a few years ago, funded [the] new [Research Centre for Stronger] Communities. An extraordinary title for a start! That was at the time that ... [an] expert at that same university was providing pro bono assistance to the community .... So you've got the community being assisted by someone at a local university who has got enough knowledge to assist them and all of a sudden that same university gets a new Chair and one can't help but wonder whether there might not follow some pressure within the university internally not to rock the boat and the pro bono work for one small community and jeopardise millions of dollars of funding. That is just one example, and one could come up with a dozen examples of how Alcoa has used its enormous wealth*

> *as one of the world's largest mining companies, to silence opposition. (former state politician)*

In Chapter 6 we analyse more closely the power dynamics underlying perceptions of corporate 'white-anting'. However, we can summarise at this stage that the practice of university funding and commissioning of research, which is mirrored internationally [62], led many to conclude that in Western Australia '[i]t is ... very hard to find any really competent ... [people] who are not in some way economically connected with the company' (state parliamentarian). In other words, Alcoa was seen to be in almost complete control of the science.

The revolving doors between industry and government were seen by outsiders as a deliberate corporate strategy to keep the regulator at bay. As one political commentator explains:

> *These are the people, you know, you've got this ... what do they call it ... the revolving door; one minute they're working for the regulator, the next minute they're out in industry telling industry how to get through the regulatory process as quickly, easily and as cheaply as possible.*

A state parliamentarian suggested that, 'the head-hunting of staff out of the Department of Environment' was a 'deliberate policy by the companies to try and capture the department's monitoring capacity.' As a result of what was described as a 'considerable flow of staff between industry and the department,' there was seen to be a 'cultural sort of familiarity', 'a pretty cosy relationship ... between the departments and certainly the bigger players.' There was a pro-industry culture that was seen to 'run ... right through from the departments right up to the Ministers', resulting in a lack of 'backup ... further up the line to actually prosecute' companies for environmental breaches. The 'exchange of staff between the department and the companies' was seen 'to soften the ... attitude.'

3. Handling dissent

Anecdotes abounded within political and departmental circles about what were seen as rather heavy-handed tactics by Alcoa to ensure the company's operations remain unaffected and its reputation untarnished. Alcoa was described by different state politicians as a company that 'likes to stay in control of the situation' and its approach to dealing with problems as 'very sophisticated but cynical'. As summarised by one politician:

They've got a nice corporate image but they've got a fairly potentially intimidating underbelly.

This underbelly earned the company the nickname 'Pittsburgh Mafia', describing a company you can 'always feel' and you sense is 'always there', ready to silence dissent and opposition (state parliamentarian). The company's treatment of the media shall serve here as one example of how Alcoa dealt with unwelcome publicity. Michael Southwell, former reporter for the *West Australian* newspaper, reported over a number of years on the Wagerup conflict, which won him the Muster Prize for Independent Journalism and the Walkley Award for News. His work, which reportedly he was pressured by Alcoa to drop, was praised as 'a series of lone-hand reporting that relentlessly pursued allegations against a very big corporate citizen' [63, p. 6]. The company's brush with the local media is told below through the eyes of a political commentator and a former state politician.

They'll use any tactic to get their way. I've never heard of a company in WA trying to bully a newspaper like they did ... they play very strategically in terms of:

> 'Well, we're going to do this and we're going to use every tactic at our disposal and if we run into bad press, then we'll use our legal armoury; we'll start off by using our charm, we'll take the Editor out to dinner, we'll take him to the ballet, and we'll take him to the theatre.' ... Editor[s] ha[ve] been offered tickets to corporate functions and what have you. 'We'll take him out to lunch, charm him, we'll try to convince him that we are just here to do some good in the community and then we'll tell him that it's really unfair what [their] reporters have been writing and maybe they haven't been getting their facts right on everything.'
>
> [An] example of the tactics of Alcoa was their attitude to Michael Southwell and [his] coverage of the issue and basically, in the end, that caused Michael Southwell to leave the West Australian, despite having won the Walkley Award, which is the top journalist's award in Australia, for his Alcoa stories. The West Australian agreed to a regime where the company was consulted about all stories before they were finalised and put in the paper and that was an impossible situation for Southwell to have the company vetting his stories, so he left. That's another example of Alcoa finding ways to silence dissent, effective dissent.

LEGALITIES AND ETHICS

> Certainly business entities have a legal obligation to be seen to make profits for their shareholders but they also have legal obligations in terms of occupational health and safety and environmental management which are just as binding. Of course it's up to the government of the day and their enforcement agencies

to ensure that the standards are high enough and that they're enforced. It's very clear and increasingly clear that some of the enforcement aspects are inadequate. (state politician)

At the community level as well as at the political and departmental levels, we found widespread perceptions that successive governments 'blatantly accepted the fact that [Alcoa] knows what it's doing.' The approach was likened to 'throwing the precautionary principle ... out the window' in that the regulatory bar was believed to have been set too low and that industry was given free reign because it was 'producing an economic surplus ... for the government.'

The Australian Intergovernmental Agreement on the Environment (Council of Australian Governments 1992) defines the precautionary principle as follows:

Where there are threats of serious or irreversible environmental damage, lack of full scientific certainty should not be used as a reason for postponing measures to prevent environmental degradation. In the application of the precautionary principle, public and private decisions should be guided by:

(i) careful evaluation to avoid, wherever practicable, serious or irreversible damage to the environment; and

(ii) an assessment of the risk-weighted consequences of various options.

Alcoa's actions might have been legal but they were seen by many observers to 'have lacked honesty and integrity'. The company's activities were perceived to have been protected and in part facilitated by the state government. Alcoa was regarded as 'a valuable industry partner providing jobs and providing wealth to the state's economy' (government department staff). Admittedly, as one community consultant observed, 'there are benefits but you have to be honest about what the impacts are ... Unfortunately that limits the profit, and it seems nobody wants to limit the profit.' A state politician concluded: it is this 'lack of integrity' which arguably shaped the nature of the Wagerup conflict 'that le[ft] ... [the Yarloop] community stranded, still sick, still disaffected, still feeling very angry and bitter, and it's all totally ... understandable.'

CHAPTER 6
SUSTAINABILITY DOUBLESPEAK

Go up Hoffman Road off the highway, there's a lookout there, have a look at what the men on the land are doing on their side of the fence, take a photo of that; then take a photo of what Alcoa has done to some of the most prime agricultural land in the state.

The people who permitted [the residue ponds] to go ahead should be trialled for murder because if that caustic gets in our waterways it's going to kill our children in the years to come. In my opinion anyone who kills human beings should be trialled for murder.

This isn't murder in the short term, this is murder in the long term; our future generations and our children that are coming on.

The residue ponds are living proof of what our government and Alcoa have done to this local community. (Yarloop resident)

We have seen how Alcoa's use of 'scientific' knowledge proved effective in legitimating its commercial and public-relations decisions. Here, we show how Alcoa worked strategically to maintain its commercial interests across two other sites of power – sites that fell beyond the legislative permission to pursue its commercial interests as established by a Western Australian Act of Parliament. The first site is the slippage between the public and private: Alcoa acknowledging mistakes and responsibility in private spaces ('off the record') versus the public image it maintained of causing no harm and creating only good outcomes for Yarloop. The second is Alcoa's centralised control of operations and public relations which undermined the authority of Wagerup managers and constrained their ability to deviate from the company line in local negotiations. We have seen how some Alcoa management personnel at the local level expressed dismay and regret for the impacts of company decisions on the communities.

These less direct ways of exercising power involve domination by those at the 'centre' – the executive management team of Alcoa World Alumina Australia – over the people and places at the 'periphery' [1; 2]. While the centralised power base of Alcoa controlled its own periphery site at Wagerup, in the process it has impacted on the Yarloop area. Yarloop is further disadvantaged in this unequal play of power due to its peripheral location to Perth [3], Western Australia's centre of legislative power. (In the final chapter of this book, we show nonetheless how the Yarloop community made itself a centre in the effort [4] to protect its interests and rights.)

These alternative power sites are evident in the political football of sustainability. Local stories presented earlier pointed to collusion between Alcoa and the WA state government. The judgement of the State Ombudsman implies the same thing.

> *The State Ombudsman has urged WA's environmental watchdog to lift its game when dealing with complaints about pollution from Alcoa's controversial Wagerup refinery ... he found the most common reply from Alcoa to the DEC when queried on the cause of community complaints was a 'no comment' response. Blank answers were also accepted by the department without further investigation ... CAPS was also upset that Alcoa had been allowed to 'self monitor' its emissions ... [in turn] Alcoa has consistently argued that the refinery meets the most stringent health and environmental standards in the world [5].*

The irrationality of the public claims [6] made by the government and Alcoa as their respective practices are questioned is prodigious. Neither party considers the absence of independent monitoring of air quality around Wagerup unusual or inappropriate. Neither acknowledges responsibility for complaints from affected residents. This refusal to accept responsibility has become a theme of this book and collectively represents the loss of capacity to protect the public's interests. Matters of natural and social justice [7] are given no due credence, while matters of economic development are upheld as inevitable [8] by both the government and the company.

In this chapter we present a critical analysis of the contradictory movements of Alcoa's public claims of sustainability and its local actions. We propose that the collusion between Alcoa and the state government was made politically palatable through corporate communications such as Alcoa's *Sustainability 08* report [9], and we examine this report for the purpose of deconstructing Alcoa's conception of corporate sustainability and its self-proclaimed role as a world leader in the area. Any claim of leading the world in sustainability is liable to be judged as baseless if made by a corporate entity that acts in contempt of local communities, but in any event, Alcoa's definition of sustainability falls far short of internationally accepted definitions [10, p. 18], and its relative absence of focus on *social* sustainability may well underpin the intractability of the Wagerup conflict [11].

INCOMPATIBLE LAND USE UNDERMINES SUSTAINABILITY

Though corporations enjoy many 'human rights', they are not required to abide by human responsibilities ... lacking the sort of physical, organic reality that characterises human beings this entity, this concept,

> *this collection of paperwork called a corporation is not capable of feelings such as shame and remorse. Instead corporations behave according to their own unique system of standards, rules, forms and objectives ... The most basic rule of corporate behaviour is that it must show a profit over time [12].*

The land management aspect of the conflict reveals the danger of leaving Alcoa to define what counts as environmental sustainability and desirable land use in the Wagerup area. In the course of researching this factor, it was perceived by community members that the ECU team itself became compromised by Alcoa's control of the study at critical points.

Too late for the many residents who felt pressured to take up Alcoa's offer to purchase their properties [13], the Parliamentary Inquiry stated [14, p. 267]:

> *The Committee finds it inappropriate that the complex competing land uses at Wagerup, Yarloop and Hamel and the strategies used to resolve them have been left to Alcoa to settle. The Committee considers that this is more properly the role of Government.*

To the present time, there is a profound silence regarding the corporation's highly vested interest in the purchase of private properties around the refinery at Wagerup. This has been conducted as a private matter to establish an economic zone around the refinery [15], suiting Alcoa's commercial and financial interests. The strategy was begun at the same time as Alcoa sought to reassure the community of Yarloop that the company did not want to destroy the town.

Other silences raise further questions about Alcoa's claims of being a world leader in environmental sustainability. The public assets of jarrah forests, lakes, streams, landscapes and atmosphere are used by Alcoa for producing alumina and profits. Previously productive farmlands are being replaced by (arguably) toxic 'mudlakes' as the residue from the refinery operations continues to encroach on the landscape. One manager told an ECU researcher off the record that the pollution from the mudlakes was by far a bigger issue than air pollution from the refinery tall stacks, and not as amenable to solution.

As one local property owner told us:

No one wants to see Alcoa close down because they employ too many people but they can't continue to run the way they are, destroying everything. It's not only the people, it's the environmental issues. The amount of groundwater contamination ... the amount of contamination in people's rainwater tanks, the amount of pollution with these two co-generator plants, acidic rain they're going to put into the atmosphere. It's supposed to be 500,000 tonnes maximum; it's going to go to something like 1.2 million tonnes a year. How can a government body allow that to happen when they've got the conditions that 'you won't exceed that'? They're going to more than double it. Someone's not doing their homework right or shall we say there's been some backhanders somewhere.

Another concerned resident remarked:

I'd like to see Alcoa go away but that's not going to happen, okay? I'd like to see them really fix their problems such as their liquor burners and the emissions. That's got to be fixed and also the caustic ponds. I hear stories that birds land in there ... ducks and wildlife and they die a pretty horrific death slowly getting burnt away in those caustic waters ... it's pretty cruel. I hear they are going to have a lot more of those

> *caustic ponds; what's that do to the environment, for the wildlife, what about that side of things? We're talking about 3000 acres of good prime land that they are going to dig up and just put caustic water in. Being a farmer I think that's a waste; that hurts.*

The key themes in these stories [16] reflect the non-commercial, non-monetary meaning of their homes and their place/community, and the deeply felt loss in having to sell to Alcoa and leave Yarloop. Many of the people affected made a community-level attempt to address the issues with Alcoa to ensure the conflict did not get stuck at the level of private, personal troubles. They also made the issue public because they thought it should concern the broader Australian population and, indeed, the global community, that disproportionate costs were being borne by local people and places in the name of progress for the greater good [17].

IDEAS SHAPING OUR ANALYSIS

We define sustainability in terms of three distinct capacities. The first is that conflict resolution practices and decisions need to follow natural justice principles and be socially just in outcomes, especially for the most vulnerable parties [18; 11]. Second, sustainability is also about the present and future safety, health and wellbeing of communities, ecosystems and landscapes [19]. Third, sustainability for industries is about being able to make profits, have secure access to required natural resources and be conflict-free with its neighbours and sponsoring government authorities [20]. Each one of these dimensions is problematic in the Wagerup case. For Alcoa to argue in its 2008 report that it is a leading example of a company that practices sustainability flies in the face of the local evidence.

In a discussion about whether Alcoa could be considered central to regional sustainability, one Alcoa manager conceded:

I think Alcoa is, but what we got into from my perspective pretty heavily, was what the old saying tells us: 'self-praise is no recommendation'. I think we got pretty heavily into self-praise; success was: negative stories = 3; neutral stories = 5; positive stories = 50. Easy scorecard. So that was focusing resources on managing the image rather than resolving the problem.

Examples of international achievements are presented in Alcoa's sustainability report.

In 2008, Alcoa was named as one of the most sustainable corporations in the world in the fourth annual Global 100 ranking of the top role-models in sustainable business practices, at the World Economic Forum in Davos, Switzerland. Alcoa has been on the list every year since the ranking was started in 2005. Alcoa was named number one in the metals category of Fortune 500's World's Most Admired Companies – this is up from its number four place in 2007 [9, p. 48].

Alcoa's achievements in these contexts lose some of their shine when juxtaposed with the realities of winners and losers at Wagerup. While we heard some critiques of Alcoa's actions internally, these were not heard in the public sphere. The perceptions of how Alcoa handled the land managment issue are described here by one Alcoa manager.

As a consequence of Alcoa's strategy of buying up properties I think we created a dynamic of some people being in it for the money. So Alcoa deliberately went out to get support for ourselves as well, so what I see that doing is walking away from the issue that it is about relationships and in fact making it worse by setting people in the town against one another, which I don't think is productive. The big issue out of all of this for me is you've really got winners and losers and a lot of losers who I think don't have – well it's a bit trite to say they're not powerful because some of the people who were involved in this and the usage of the newspapers and so forth was a pretty empowering process for them, but I think you've got a lot of people hurt by what's happened and that's what I'd like to see not happen again – and not necessarily from the fact of hurt by the emissions but hurt by the way the whole thing has been handled ... there's going to be better ways of handling it more effectively into the future.

The sense of belonging to place is an important aspect in accounts of environmental sustainability and social justice [21]. No number of heart-wrenching stories by locals has shifted Alcoa managers to accept the collective validity of people's loss and harm as a result of Alcoa's actions, despite the fact that some of the impacts detailed in this book have been acknowledged in the public domain. The company has persisted in its practices of securing private land and, from 2004, took steps to secure an expansion of production even as this was commonly perceived as an enormous threat to an already collapsed community and town.

This dogged pursuit of commercial interests represents a blatant use of power that is only minimally constrained by government intervention.

When the government attempted to respond with the Yarloop Sustainability Plan to help ensure Yarloop's survival as a town, locals perceived that Alcoa used this to sidestep putting its final seal of approval on a ten-month-long negotiation with residents on fairer land management rules.

This dominance of corporate power requires a more developed understanding of sustainability, incorporating a critical analysis of power abuses, if the government and Alcoa are to be successfully held accountable. With this intent, we closely examine the land management meetings and related company–community conversations instigated by the ECU research team in 2002, employing 'phronetic social research', an approach well suited to clarifying political situations through the use of detailed stories about 'who's doing what to whom' [22, p. 140].

The phronetic approach calls for an interest in the way power is defined by those who exercise it and in how 'important' knowledge is used and constructed, recognising that 'knowledge ... can be marginalised by power and power ... can produce knowledge which serves its own purposes best' [22, p. 142]. Drawing on these ideas we provide further evidence of how Alcoa has, in almost all instances we have studied, maintained control over what counts as knowledge and over what needs to be done about any issues it defines as problems. Of interest are the observable slippages, what we call 'doublespeak' or contradictory movements of power which, when seen through the phronetic research lens [6; 22], illustrate how the exercise of unfettered corporate power has been possible.

UNPACKING THE DOMINANT STORY-LINE

Alcoa and the state government have left traces of evidence of the exercise of power and knowledge along a dominant story-line of advantaging commercial and economic interests. That the public at large has not taken up the full extent of its democratic rights and responsibilities may be due to a widespread assumption of the inevitability of corporate power of this magnitude [23], a point taken up in the final chapter.

In its formal sustainability report [9], Alcoa draws on a wealth of dedicated resources, listing achievements recognised by some of the most prestigious entities in the world, and claiming validation of its practices in published research by academic and other skilled experts. The strategy undertaken by the ECU research team of validating local knowledge and experiences [24] unsettles these public claims. The research discerns a corporate doublespeak that presents information used to protect and promote commercial values and interests in the guise of legitimate scientific truths. This corporate politics is known intimately as a reality by many of the locals we interviewed. They understand that Alcoa's commercial interests are served through its sustainability reports and media releases whose dominant story-line attempts to hide the politics of corporate practices that can only be considered sustainable in the narrowest definitions of the term.

In these final two chapters we employ a research schema [based on 6; 22] to assist us in unpacking the nature of the power politics of the Wagerup conflict. The schema asks the following 'value-rational' questions which we have adapted to the key parameters of the Wagerup conflict [22, p. 145].

1. Where is sustainability going in the Wagerup situation?
2. Who gains, who loses, and by which mechanisms of power?
3. Is it desirable?
4. What should be done?

The first two questions are addressed in this chapter and Chapter 7 shows why we think it undesirable that Alcoa's definition and practice of sustainability are left unchallenged. (The absence of any challenge is as much by default in an insufficiently robust democracy with weak legislative controls as by deliberately ignoring the lack of substance to Alcoa's social sustainability efforts.

WHERE SUSTAINABILITY IS HEADED AT WAGERUP

According to Alcoa's 2008 sustainability report [9, p. 5]:

At Alcoa corporate social responsibility is ingrained in our business, it's part of our DNA. Our sustainability embraces both our own operations and also the communities in which we operate.

Yarloop in 2009 is not the Yarloop of 2002 the ECU research team encountered when they witnessed first-hand the unexpected substantial collapse of the town as many property owners sold to Alcoa and moved away, taking their social and financial assets with them. In a video made by the research team at the time, an Alcoa representative said that it was not acceptable for there to be such incompatible land uses – that private property owners were perceived as incompatible to Alcoa's commercial needs. Alcoa created its land management proposal without local or state government approval, which many locals regarded as a very clear example of the unfettered exercise of corporate power. A resident interviewed for the video said it was like Alcoa had 'drawn a yellow line of pus down the middle of the town,' while others claimed this was a disaster for the sustainability of the town [13]. In a separate interview another resident remarked:

The local community has basically been wiped out by their buy-out program, which they didn't put a lot of thought or planning into. They've changed it so many times since its initial inception. I think they're reactionary to their issues rather than being proactive. They get away with whatever they can until they have to do something about it, rather than trying to be proactive about it.

Alcoa's idea of sustainability was not embracing the company's 'own operations and the communities' in which they operated. The following excerpt from an interview with two Yarloop residents, fictitiously named here Tom and Sally Townsend, is a testimony to the misfit between corporate interests and community interests, which is a signifier of unsustainability [25] as well as personal injustices.

> *Tom: Well Alcoa did the real smart move, that being the old divide and conquer. They had the A and B zone and they just ran it right down the middle of town. That was the beginning of the end for Yarloop because you had one side that was going to get 35% and the other one was going to get nothing. So you've got one bloke on one side of the street saying why the hell do you get 35% more than I? It was all arguing: why should you get this and I'm not getting that. That wasn't very good. That wasn't the government; it was Alcoa doing that.*
>
> *Tom: Everything you've worked for all your life you lost.*
>
> *Sally: Oh, it was dreadful. It was just dreadful. I felt like blowing the place up ... blowing Alcoa up. And you couldn't do anything about it. It was like fighting.*
>
> *Tom: They used to play good guy, bad guy all the time ... One bloke would be the good guy and the next one down would be an absolute arsehole ... It's just a bloody game with them. That's all it is, just a game.*

> Sally: I thought I was going to have a heart attack before the end, I really thought, I can't live with this any more. I can't, it was so emotional. Very traumatic.
>
> Tom: To lose the farm, even though, yes, we got a fair few dollars, no doubt about that, but we lost something that I thought I'd always have. And when that is passed down to the next generation and the next one if they want it. You just couldn't stay. Well you could stay there all right but your life expectancy certainly would have been shortened in my opinion.
>
> Sally: Then there was seeing everybody go. That was strange. Hanging on and hanging on and hanging on to try to get a fair deal ... because it was Tom and your livelihood, and I had to leave my job. I worked in Rockingham and I had to leave my job as well. Then I worked in Bunbury and it was all very hard, wasn't it? Then seeing friends go and houses becoming derelict.
>
> Tom: When Alcoa first came there they pushed down all the houses that were on the farms; they only left a few.
>
> Sally: We were just watching it all.

Alcoa, by these accounts, in pursuing its land-use interests divided a town, precipitating a fear-based domino effect of residents selling to Alcoa and leaving the area. Sally and Tom's story was mirrored in other residents' experiences and became a public issue with social justice concerns [26]. In stark contrast, in *Sustainability 08* Alcoa describes its philanthropy in the Yarloop district in a positive, rational manner as an example of community partnering. The account somehow manages to avoid any mention of the adverse effects of its unilateral exercise of power.

> *We also continue to provide ongoing benefits to communities in the area through the Wagerup Sustainability Fund ... The fund acts as a catalyst for building partnerships and programs with other organisations essential for the vibrancy and sustainability of surrounding communities by supporting social, economic, and environmental projects that promote the growth of the region. Specifically, we have linked our annual donation to the fund to production at the refinery. Therefore, when we benefit from production improvements, the community and the region will directly benefit, too [9].*

There is no comparison between this Fund as an egalitarian resource building and sharing initiative and the working out of the land management strategy, which was named by a consulting town planner as the establishment of an economic zone of influence by Alcoa with financial imperatives [27]. In return for the power to orchestrate the purchasing of what some residents believe is more than half of the town of Yarloop, Alcoa gave Yarloop an opportunity to apply for Fund monies. The playground in the main street of Yarloop was funded by Alcoa but for a time was rarely seen to have children playing in it. It is at this social and community level that the extent of depletion of social capital and destruction of a town is most evident.

A MORE ROBUST AND SOCIALLY JUST IDEA OF SUSTAINABILITY

When land use is disputed, sustainability principles would suggest delivering a power sharing process and a carefully negotiated agreement between Alcoa and the communities adjacent to the refinery. In turn this would be sanctioned and facilitated by government, not by an Alcoa-designed

Community consultation (Photo by H. Seiver)

set of rules. The reality was far more contested than this, as illustrated by a local resident expressing his disgust to an Alcoa manager who he saw as blaming the community for the problems.

> *Well, if you were a good neighbour and a good corporate citizen that you make yourself out to be when you hand out all these awards and you put your name up everywhere, wouldn't you have just done that? Why did we have to force you to do it?*

The resident continued by saying to the interviewer:

> *.... And it virtually was that scenario; it was only when they started getting really bad press that they started doing anything and that was when the DEP got serious too. So you've got two issues: you've got a reticent company; [and] you've got a recalcitrant DEP, or whatever they call themselves now, that weren't really interested either until we started giving them a hard time in the media and they were getting*

> *bad press, then they got serious about it as well. So you've got two problems; you've got a government department, or actually it was even then the Health Department weren't that interested; they came on board a lot quicker than the DEP but in the beginning the hierarchy there were willing to fob us off as well.*

Edith Cowan University began working with Alcoa Wagerup in June 2002 with the broad brief of enabling constructive relationships between local leaders and residents and Alcoa, and finding workable strategies to the shared issues. From the outset the fact that we were funded by Alcoa brought a tension to any attempts to build trust across the divide between the town and the mining company. We tried to address this by creating spaces where the conflicted parties could meet and be supported to find their own solutions through dialogue [28; 29; 30], believing it was appropriate for Alcoa to be seen to be paying for such expertise. We remained very alert to the risk of colluding with Alcoa and for this reason insisted that all discussions with Alcoa and any publications about the research by the ECU research team were to be in the public domain for all to access. As one of the locals recalled some time later:

> *Some of us used to go to the land management meetings. That was for ECU and Dyann Ross. Craig went for over 12 months and in the end he'd had enough; he couldn't take the crap anymore and Craig and I had always tagged one another.*
>
> *Then they chopped Dyann Ross's feet off and we told Dyann that when she first started: 'Alcoa will only let you go so far and then you're finished.'*

For a period of time, though, the active presence of a third party [31] was useful in identifying and documenting the unfolding events and impacts, particularly relating to the land management strategy. Concerns and fears about health when compounded with perceived threats to property values became a potent recipe for mistrust, misinformation, outrage and panic. ECU's early contact with the community revealed heightened and progressively worsening fears about Alcoa's intentions, which in their most extreme were expressed as Alcoa deliberately attempting to shut down the whole town. This was partly based on the fact that an unexpected number of residents had sold to Alcoa, including residents in the older part of Yarloop, which was not in the proposed 'buffer' area.

In the ten months of meetings between Alcoa and property owners from Yarloop and Hamel, an enormous range and complexity of issues were identified and discussed. In some cases, issues were advanced to recommendations based on quite sophisticated problem-solving and cooperation across different viewpoints. The meetings represented one of the few open, ongoing public forums where a dialogue was established between

Alcoa-Yarloop land management meeting (Photo by H. Seiver)

the parties, and occurred during a period of significant upheaval and conflict. It was perhaps inevitable that the strength of the process was also its point of vulnerability insofar as the land management meetings became the location for attempts to find solutions for all aspects of the interconnected issues.

WINNING AND LOSING AND THE MECHANISMS OF POWER

The big issue out of all this for me is you've really got winners and losers and a lot of losers who have ... been hurt by what's happened and that's what I'd like to see not happen again – and not necessarily from the fact of hurt by the emissions but hurt by the way the whole thing has been handled. (Alcoa manager)

When the ECU research team spoke to community participants in 2002 and 2003 they were told of a number of unresolved issues regarded as crucially important to the viability of the towns and the fair treatment of those affected by Alcoa's Wagerup operation. Some of these issues are presented here in note form to show the methods of unpacking their various elements.

1. There was continuing disquiet expressed regarding health impacts, in some instances these affecting people further afield, attributed to the construction of the tall stacks.

Who won?

Alcoa built the tall stacks in 2002 without an environmental and social impact study as part of a technical intervention to control its emissions. This was done in the context of national media reports of pollution from its Wagerup refinery, including noise levels above the legally permitted levels. To a naive onlooker this gave the impression of being a responsive and responsible neighbour to surrounding communities.

Who lost?

With the facility upgrade it then became even harder for affected local people to be heard because there was concrete evidence of money spent to fix the problem, with two very imposing tall stacks which could be seen from miles away as practical proof of Alcoa's efforts.

Now there was a new category of people affected: those who could see the tall stacks. Previously the refinery infrastructure was not visible in an otherwise rural landscape.

Residents previously affected by refinery emissions and new complainants in Cookernup and outlying farms complained that the tall stacks made the emissions worse because they spread emissions further.

How?

Complaints about more people being affected were discounted by Alcoa personnel in public meetings. Alcoa representatives were able to avoid responsibility for any malodour, noise, and increased visual pollution while simultaneously acting to improve its emissions controls, this facility upgrade in itself implying there was room for improvement.

There was no question that the tall stacks should remain, much as the refinery is presumed to have to stay. Occupying place (the refinery site), intruding into airspace, as part of undertaking significant facility changes without consultation or government approval were related strategies that allow the contradictory positioning assumed by Alcoa – 'there is no problem, we have just fixed the problem'.

2. People questioned the wisdom of trying to solve land management issues with the uncertainty and fears still being expressed that emission controls had not fixed 'the problem'.

3. The need for a 'buffer' remained a big question from the outset of the meetings.

Who won?

Until recently, Alcoa has continued to buy up private properties in the area surrounding its Wagerup refinery as part of securing control over land use outside its legally approved refinery and mine-site footprint. Alcoa personnel claimed in public meetings that incompatible land usage was not in their commercial interests and they now own or have bought a significant part of the Yarloop townsite and district.

Some property owners who sold to Alcoa report that they were pleased with the financial settlement gained by selling to Alcoa.

Some people who wanted to leave Yarloop who didn't have health and other issues related to Alcoa's operations were able to sell to Alcoa.

Who lost?

Western Australia's democratic process lost as noted in the Parliamentary Inquiry [see 14] because of the failure of Alcoa to follow due process for obtaining a state approved industrial zone around its refinery.

Individual property owners, including some business owners lost in the context of perceived health and social impacts in the community. Many people told ECU researchers that people sold to Alcoa under some degree of duress.

Whole towns were subsequently impacted by the

domino effect of people selling and leaving as their neighbours, family and friends left and fears about the loss of vital infrastructure heightened.

The district is losing as Alcoa's industrial footprint continues to take over prime farming land and related employment opportunities.

Non-property owners lost as they have tended to not be seen as part of the stakeholder group who might make claims of impact against Alcoa.

The traditional owners of the land lost as they have not been acknowledged or consulted.

Social justice lost as the bigger issues of compensation for losses and lack of accountability of the government and the company have, to date, not been addressed to the satisfaction of residents and CAPS.

How?

The language of calling the creation of an 'economic zone' [15] around the refinery a 'buffer' is in itself misrepresentative. While challenged by locals, the continued use of the word (including in this book) created a sense that the land use in the adjacent community was a problem, according to Alcoa, and of Alcoa needing to 'buffer' itself against the town's encroachment on its commercial interests.

All this has continued to occur without formal state government approval to do so and outside the usual town planning processes.

Alcoa was often challenged in the meetings as to the need for creating a buffer. Complainants were dismissed but not convinced by the incompatible land use argument. As the social impacts became evident there was an attempt by Alcoa to avoid responding to property owners outside the areas it considered eligible, but not to stop its drive to buy up properties or revert to a formal application for a buffer. Thus a 'push on' strategy in spite of opposition and evidence of harm being caused.

Individual private commercial transactions were the mechanism by which Alcoa gained control over large amounts of property in the Yarloop district.

The Federal Act, which provides guidelines to overseas companies regarding purchasing property, has not contained this exercise of corporate power.

The adoption of different buyout rules for property owners in different locations had the effect of setting people against each other, further fracturing the community and making a concerted collective resistance to the buy up impossible.

Alcoa's refusal to link the buy-up to an acknowledgement of health and social impacts allowed a deep contradiction to persist – providing an ostensibly generous and helpful offer to purchase properties in the event people did want to leave Yarloop, but not recognising the legitimacy of many people's health concerns.

This dynamic became further entrenched with the state government requirement that Alcoa provide a supplementary land management offer to property owners outside the original areas if they could prove they had been affected by Alcoa's operations. Here the government directly implicates itself in the mechanisms of power operating under Alcoa's efforts to secure its own personal buffer.

4. The rights of property owners and the towns included in the proposed 'buffer' could not be sufficiently clarified or assured.

5. A social impact assessment and social mitigation

package was not implemented before the land management offer by Alcoa.

6. People spoke of how this has placed an onerous burden on them to name and seek adequate responses to the issues as they were being affected by them.

Who won?

In an uneven power setting, not sufficiently mediated by authorised government representatives, it tended to be Alcoa who won.

Community members who did not challenge Alcoa or who kept quiet during the controversy were able to protect themselves from this onerous burden.

Community members who held some favour with Alcoa or who publicly sided with Alcoa were not required to name the issues.

Who lost?

Community members who felt impacted and who spoke up about the issues, especially those perceived as being leaders, lost.

Affected residents had to defend their experiences of health and other impacts in a context where this was not welcomed and not seen to be a problem.

People who regularly attended the community meetings often did so at considerable personal and family cost.

Two people who were integral in raising the issues and trying to get Alcoa to do something to fix the problems died suddenly. Many locals told ECU researchers they believed the stress was a factor in their deaths.

How?

Alcoa was advantaged by individuals and activist

groups naming the issues, as it could subsequently position itself to protect its own interests, or less often, to ameliorate the adverse effects of its actions.

Another mechanism of corporate power was the invasion of privacy whereby individuals who were challenging Alcoa in public meetings or the media, often had their personal and financial business laid bare to Alcoa.

ECU researchers were told of some individuals being discredited as a result of Alcoa personnel coming to know about their personal circumstances.

There was a perceived risk of being construed as troublemakers or anti-Alcoa if people spoke up.

There was a perceived risk of getting an unfair buyout price for property owners who were anti-Alcoa, especially if they were working for Alcoa.

7. The extent of population change was rapid and devastating for the communities affected. Many perceived this as a deliberate strategy by Alcoa to close down the town.

8. The population increases that might have been expected in the towns have not occurred since Alcoa started buying properties (in early 2003).

9. There has been an expression of the urgency to address the impact of these changes on the small businesses and other income earners who lost work with the cessation of development in the towns.

10 The boundaries of the designated Area A have been consistently contested, with the most common arguments being that it should be extended to include the whole town, or alternatively, that it should not exist at all and Alcoa should compensate any affected property/individual.

11. For the short term, community participants argued that priority should be given to ensuring fair treatment of the people who felt they needed to leave, even though this has a flow-on effect threatening the viability of the town.

12. Those who wanted to stay needed guarantees for their health and safety, and security for their property values.

13. It was often argued that there should be no difference in how people were treated – thus all purchases by Alcoa should be in accordance with the Area A offer and should include compensation for losses relating to social amenity and personal harm.

14. The state and local governments' roles and responsibilities in the issues were consistently questioned, with dissatisfaction being the predominant sentiment.

15. The government sustainability package for Yarloop was noted but the loss of beds at the Yarloop Hospital was considered by many to reflect a lack of commitment to their promise to protect crucial infrastructure.

16. There was a strong perception that any expansion or increased production at the Wagerup refinery would further adversely impact on Yarloop and Hamel.

17. Dissatisfaction was felt throughout the meetings due to the belief that Alcoa set the parameters of what could be negotiated and didn't substantially shift from these.

Meanwhile, Alcoa still claims an excellent record in the face of unresolved social justice in its dealing with local communities throughout Australia [9]:

Specifically, [Alcoa's community] survey focused on perceptions of our performance and reputation, which are strongly driven by the extent stakeholders perceived they experienced both just treatment in their interactions with Alcoa and high quality relationships.

> *In this study, just treatment was defined as stakeholder perceptions of ethical and consistent decision making and feeling respected and satisfied with Alcoa's explanations.*
>
> *The quality of our stakeholder relationships was measured by assessing the level of social capital in the relationships between stakeholders and Alcoa and between the stakeholders themselves. Social capital is described as the trust, mutual understanding and shared values between members of the community and the company that enable them to work together for shared outcomes.*

The report highlighted Alcoa's key results [9, p. 19]:

- *Alcoa's overall performance is viewed positively, with the lowest average ratings of at least 3 out of 5;*
- *Alcoa's reputation is high across most sites with the exception of Wagerup, which was the only site to rate less than 3 out of 5;*
- *Stakeholders felt Alcoa treated them justly, with scores of at least 3.5 out of 5;*
- *The satisfaction of stakeholders in regard to their relationship with Alcoa is also generally high, with most sites rating more than 4 out of 5.*

The overall rating for each operational site was as follows:

Point Henry 4.35
Anglesea 4.03
Portland 3.97
Pinjarra 3.73
Kwinana 3.68
Yennora 3.5
Mining 3.44
Wagerup 3.23

In defence of the low score received at Wagerup, Alcoa [9] writes:

> *Results from the site-level analyses led to recommendations specific to each site. Most are aimed at strengthening already good relations. At Wagerup, however, we recognise there is a need to increase our communication and improve our relationships with some stakeholders.*

This neutral representation of a long-running complex controversy allows no confidence that Alcoa has learnt from its experiences.

CONTROLLING THE INDEPENDENT RESEARCH AGENDA

> *The issue is to ensure, absolutely, that the weight of control and key determinants of outcomes is either adequately balanced between Alcoa and the community or, if anything, weighted in favour of the community. If excessive control rests with Alcoa (which may be a natural and preferred tendency for Alcoa in the absence of advice to the contrary) and there is any room whatsoever for criticism of dissatisfaction with the process or its outcomes, the process will run the risk of failing, and ongoing controversy is likely to be experienced [32, p. 7].*

There were perceptions in the community that the ECU researchers, who were on the ground at a critical time in the deterioration of relations bettween Alcoa and neighbouring towns, were thwarted in their attempts to name social justice issues, point out the lack of natural justice in how decisions were made, and reduce the ongoing resistance to their brief as researchers. For example:

> *Mr Olney said Alcoa had made sincere efforts in responding to concerns about health, social and*

community impacts related to the Wagerup refinery and Alcoa's land management strategy ... Key Alcoa initiatives for the sustainability of Yarloop and Hamel (include the) ... ECU project – providing a full-time expert facilitator for on-ground support and community drop-in centre [33].

The research was presented as providing a service to help the community cope, without mention of the confronting and demanding meetings which were so challenging to Alcoa. It was closer to the truth that ECU researchers, while engaging people and showing care and respect, worked really hard to convince people that it was worth giving Alcoa a chance to make amends and that it was worth filling out surveys, coming to the meetings and putting in hundreds of hours of unpaid time to help Alcoa.

The behind-the-scenes meetings with Alcoa personnel often involved frank and sincere discussions. But more than this, much attention was given to enabling Alcoa's capacity to act with goodwill and to let go of a fairly entrenched victim mentality [34] of being under siege from troublemakers. In this regard, it was a constant ethical challenge to maintain the research partnership because of Alcoa's insistence that all actions taken, including the ECU research project, needed to serve the company's interests. It was obvious that Alcoa's good intentions were not enough – the literature explains it in the following way:

In a [corporate] structure, human morality is anomalous. Because of this double standard – one for real human beings and another for fictitious persons called corporations – we sometimes see bizarre behaviour from executives who presumably know what is right, yet behave in a contrary fashion [12, p. 289].

A stronger ethical base to its corporate actions might have helped Alcoa Wagerup in a complex conflict such as this – there was often seen to be a disjuncture between its desire to be a good neighbour and an individual personnel's actions.

An ethical company is not afraid to discuss the undiscussable, promotes a culture of openness, trust, dialogue and disclosure ... It recognises that fundamental human values of social and environmental responsibility, community commitment and dignity can sit alongside profit maximisation. [35, p. 195]

Early in the ECU's research work with Alcoa, the company secured a Professorial Chair and gave its name to a new research centre (Alcoa's Centre for Stronger Communities) at Curtin University of Technology [36]. The Centre was given considerable coverage in Alcoa's media releases and promised to undertake national research, enabling its neighbouring communities to be strong and resilient. When questioned about this initiative and its relationship to the ECU project, we were told there was to be no relationship, as the new Centre was not going to be addressing the specifics of the Wagerup issue.

This development was seen by many in the community as a strategic business decision that made it possible for Alcoa to be seen to be doing something about communities but not the one that was most problematic for them. What was also troubling was the way Alcoa positioned the Centre's research priority on stronger communities as if, relative to Alcoa, communities are weak or problematic. Meanwhile, we found the Yarloop community very strong and willing to invest incredible energy, time and money into protecting their assets and people.

It was stated in the community that the ECU research was about enabling Alcoa to be a more ethical, socially just company to its immediate neighbours [37], but it was believed that Alcoa would not allow a public statement of this. Many felt the work was not so much about building the community's capacity to be sustainable but about trying to show Alcoa how its actions undermined its own values and sustainability agendas as well as the town's viability and sustainability. Alcoa to this day believes its own consultative forums and surveys for its own agendas are the proven recipe for building a community licence to operate, as below [9, p. 19]:

> *Key learnings from the survey are directing our efforts on the following:*
> - *developing an Australia-wide stakeholder engagement framework;*
> - *reviewing and refreshing community consultation networks; and*
> - *increasing stakeholder motivation to collaborate with Alcoa on shared issues.*

The land management experience suggests other potentially productive ways to meet the aggrieved community on its ground and to proceed on its terms rather than on Alcoa's definitions of the problem, scope of research and limits of its philanthropy. Many in the community believed that Alcoa was motivated to collaborate on issues that were of interest to the company but not those of interest to the community [11].

The implications of our claims here are mapped through the following historically embedded account of a critical stage in what became a failed attempt at dialogue for social and economic sustainability at Wagerup. It derives from the research team's unpublished report [15], given to the community and Alcoa at the time when the months of meetings had been halted by ECU researchers. The meetings stopped due to perceptions of transgressions by

Alcoa staff of the 'warrants' [38] which might be expected in a civil, democratic society [39; 40] to legitimate a company's or government's community engagement and attempts at problem-solving. With the transgression of these warrants the very legitimacy of the meetings was eroded. This point is illustrated in the following excerpts taken from the publicly available 2003 report to stakeholders by the ECU researchers [15].

> *There are real limits to what can be achieved through a direct participatory, collaborative approach to problem-solving as was attempted here. The power dynamics in the relationship between the parties were typical in the simplistic sense of a multinational company making decisions which have had considerable impacts on their neighbouring communities – a David and Goliath type scenario. This is the zero/sum power dynamic where one party is seen to have all the power and the other party to have no power [see 41]. Yet the power dynamics were also at the micro level much more fluid and embedded in particular relationships, at least for some of the meetings. Thus, I saw Alcoa participants allow a range of community opinions that they did not necessarily agree with – allow in the sense of respecting these as valid for the people and worthy of placing on the table for consideration.*
>
> *While there were asymmetries of power, community members were active in representing their interests and in this sense they were empowered/powerful overall throughout the collaboration. The community participants exercised a range of powerful acts from passionate arguments, demands for particular strategies or decisions re process, resistance to others' ideas and non-cooperation by withdrawing from the meetings in protest, etc. When at the table, power*

was not held exclusively by one person or party at the expense of others but was continually negotiated and grappled with as people struggled to be heard or to achieve a particular agreement.

There were brief experiences of genuine power sharing (although few tangible results for all the discussions).

When this occurred, power can be cultivated as a productive resource [see 42] and the idea of a winner and a loser was thereby unsettled as some sense of win/win for both parties was sought.

These introductory comments notwithstanding, I want to note here that the greatest point of vulnerability in the work relates to the movement from collaboration in the meetings to the executive management team of Alcoa having to respond to the recommendations. This placed the parties in the traditional win/lose power dynamic that had underpinned much of the conflict. While it was hoped that the process had been supported by the Perth-based Alcoa executive through to the fine detail of ideas, and that Alcoa Wagerup managers then had little more to do at the end than double-check with senior colleagues, the intervening lapse of a couple of months has been quite dispiriting for the community participants.

There are echoes back to the ... time where Alcoa previously consulted with the communities and then developed its 2002 Land Management Proposal. Echoes which are reverberations to the loss of sovereignty by the communities as Alcoa commenced to purchase a significant amount of the freehold real estate, changing the character of the two towns and surrounding areas. The real estate belonged to people who made up these communities in undefinable ways.

And the individual actions of these people in selling to Alcoa had the cumulative effect that perhaps no one person or Alcoa would have wished.

There is a deep and possibly irreconcilable difference of interest that has underpinned the Land Management work. Namely, that to the extent that Alcoa wishes to own and control areas of real estate that were in the hands of residents of two small communities until recently, then to a similar extent have the communities, who are the living dimension of this real estate, been disenfranchised from this resource and heritage. It is a matter of social justice that the people impacted on by Alcoa's decisions are not only treated fairly as has been the main impetus of the current revisions to the Land Management Proposal, but as well are given full control over decisions which affect them.

Where this is not achievable, a partnership that holds to open collaboration and refuses the shift to consultation or imposed decisions may foster some experience of justice that enables healing and recovery of individuals and places.

Early in September 2003, ECU believed that there remained an opportunity for the current work between Alcoa and Yarloop/Hamel to return to a genuine collaboration, despite the momentary holding of 'the power' (to make THE decision) by Alcoa. We understood that Alcoa Wagerup is very mindful of what the process has been attempting to achieve and values the civil engagement with [its] neighbours. [Alcoa] wants as much as the communities involved for this to work.

An epilogue
This epilogue has been written in early November 2003 – two months after the bulk of the report was

written and sent to stakeholders for validity checking. The intervening time has found the completion of the report held up in a context of loss of confidence by the parties. On 18 September, Alcoa informed ... community members at a specially convened land management meeting that Alcoa was unwilling to uphold the underwriting option for people in Area B. At the same time ... all the other recommendations were agreed to.

The shock and disappointment of those present soon brought the meeting to a close without any clear next step. The underwriting initiative was considered crucial by many residents, including those living in Area A as it was the best alternative that could be agreed to at the table for security for property values over time. It would also have enabled an equity with Area A property owners who have security in the form of a life of the refinery offer to sell to Alcoa.

Alcoa explained [its] decision at a community consultative meeting (i.e., not to the land management meeting participants) in the following manner:

- *[The underwriting option] was not consistent with Alcoa exiting the property market which many residents of Yarloop had indicated they wanted*
- *There was concern that in the future it could create issues for residents in particular those who were new to the area, and*
- *The amendment had broader planning implications that needed to be considered in the context of the current state government initiative [43].*

Alongside this was a strong community interest in obtaining compensation for losses related to social amenity, security for residents and equal treatment of all the townspeople (and other impacted properties)

.... The perceived unwillingness of Alcoa to continue the negotiation on the issue of security seems to have spread to a lack of trust and confidence that other complex issues can be satisfactorily addressed ...

There is a high risk that the warrants that gave legitimacy to the meetings may be broken if the parties do not attempt to retrieve the goodwill and shared common ground as a basis for regrouping. The warrants related to willingness to respect each others' points of view, to trust people's sincerity and capacity to follow through with the sentiment from the meetings, and together to authorise a genuinely collaborative partnership to address shared concerns.

The commitments embedded in the recommendations that Alcoa does agree to, hold further challenges, if the parties are willing to return to the table. Namely: how to persist in the face of threats to the collaborative process ... to generate win/win outcomes in a changing context. Should the parties not return to the table, the common ground will not be built on and the resultant loss will be quite profound for both parties.

After this report was written, the research team made a decision to discontinue attempts at enabling collaborative problem-solving [11]. Community perceptions at the time were that the decision was triggered by Alcoa management reneging on a commitment to take a key progress report on the research [15] to the community for discussion and refusing to allow it to be shared with the involved stakeholders. What was clear to all parties was that the warrants of willing attention and sincerity in the ECU–Alcoa partnership had been broken. The warrants of substantive contribution to the resolving of the problems with the affected residents was already broken and along with it went the hope that free dialogue with its neighbours was a viable and honourable way for Alcoa to address issues relating to the refinery's operations.

Under pressure from the Parliamentary Inquiry [14], Alcoa Wagerup did return to the table with some of the local people to work out a more equitable outcome along the lines considered by the land management meetings. Even then Alcoa remained adamant that the offer to purchase properties was not an admission of causing adverse health impacts in the area. Strangely, when the government intervened with its supplementary process to allow property owners outside Alcoa's A and B areas the opportunity to sell to the company, it required people to show they had been adversely impacted.

Soon after the deepening crisis in Yarloop [44], and with the Parliamentary Inquiry barely tabled in 2004, Alcoa failed to listen to the professional advice of key parties, including Professor Darcy Holman, Chair of the Medical Practitioners' Forum, to give the community time to recover. Despite constantly denying any intentions to increase refinery production, Alcoa proceeded with plans to gain approval for a major expansion of its operations. In Alcoa's 2008 sustainability report [9], the unsustainability of proceeding even when it had gained approval was couched in the following upbeat corporate speak:

> *... As part of our strategic response to the financial crisis, we postponed the expansion of our Wagerup alumina refinery in Western Australia until overall market conditions improve and there is greater security in Australian energy supply.*
>
> *Despite the current energy and economic challenges, our commitment to the strength and vitality of the communities surrounding our facilities is unfaltering. In Wagerup, for example, we anticipate the expansion*

project will ultimately move forward, as it offers a strong investment return.

The buy-up of properties continued well into 2008.

THE MECHANISMS OF DOUBLESPEAK

Unable, or unwilling, to see the significance of non-scientific knowledge as conveyed by residents and independent research efforts, Alcoa at the same time largely ignored social and place-based environmental dimensions of sustainability issues in a feast of corporate doublespeak. 'Not seeing' perhaps afforded Alcoa the comfort of not having to take any responsibility for addressing claims of harm and injustice by people of the Yarloop district. Alternatively, allowing a slippage between its highly self-defensive formal submissions and proud claims of sustainability, and its refusal to act responsibly outside its narrowly defined commercial interests in Yarloop, can be viewed as a key mechanism of power.

The Yarloop and District Concerned Residents Group made a submission to the Parliamentary Inquiry suggesting the following reason for not upholding the collaborative research work between Alcoa and Yarloop:

... continuing to refuse to address security for residents (outside Area A) further heightens concerns that Alcoa are merely ensuring that they cannot be held accountable for any increase in impacts as a result of any production increase, effectively making sure no precedent is set as a result of the current controversy and conflict ... therefore no commitments are made that Alcoa will be bound to ... [It] is the only plausible explanation as to why they have not acknowledged, let alone followed, the recommendations of the consultants engaged by ECU as part of the land management meetings. The purpose of all that work

> *was to ensure 'fairness and equity' and 'win/win' outcomes for all [14, p. 265-266].*

Here is the dynamic of corporate power refusing to accept scientific and rational/expert knowledge when it is not perceived to serve its commercial interests. Subjective factors influenced what Alcoa counted as scientific and valid knowledge upon which to base its business decisions. These subjective factors underly the deeply political nature of Alcoa's community relations and land-use strategies, rendering claims about the compatible and mutually reinforcing nature of commercial and community interests irreconcilable.

Moreover, there is no evidence that this corporate irrationality has been questioned in-house, according to the rosy depiction in Alcoa's sustainability report [9, p. 17].

> *The future of Alcoa in Australia remains bright, because we have one of the most efficient, integrated aluminium production systems in the world. We have access to excellent bauxite reserves, energy-efficient alumina refineries, and integrated aluminium smelters and rolling mills. Our employees live in the communities in which we operate, and we believe we are an integral part of those communities. We also continue to focus on managing the social, environmental, and economic aspects of our business.*

Yarloop, the place, has been traded, one private property at a time, for a mining company's gain. The context of fear, uncertainty and loss in which this occurred was created by what stikes as collusion between the state government and corporate interests and the collision between community and corporate values, knowledge bases and interests. To speak of being a sustainable company is only possible to the extent that Alcoa leaves out of its key reports the dissenting voices, unpleasant research findings and contested parts of Alcoa's Wagerup history. Also left out of any public account of its practices are the adverse social impacts that closely link with abuses of power, as seen in the land management debacle.

Sustainability, even in its narrowest of definitions, has little reality when social justice is denied to a whole town and landscape. In relation to Alcoa's commercially driven decision to pursue an informal land management strategy, it was noted at the Parliamentary Inquiry that in so doing basic rights for the public were denied.

> *Most significantly, a formal buffer provides an opportunity for the community to have input into what land uses are around the refinery and the way they are managed. This has proven very difficult with the current informal approach [O'Connor, Chief Executive Officer, Shire of Waroona cited in 14, p. 255].*

As disturbing as Alcoa's confident claims of excelling on sustainability indicators is the government's response to the land management issue with the Yarloop Sustainability Plan and its general lack of follow-through on the recommendations of the Parliamentary Inquiry [14]. Residents often told us they were not surprised by Alcoa's behaviour but they expected their government to do the right thing by them.

A deeper level of injustice emerges as the robustness of democratic processes has been found to be wanting, with both the government and the broader populace seeming powerless (or lacking the interest) to ensure that social justice and sustainability are achieved at Wagerup. We dedicate the final chapter of our book to this concern and to the efforts and influence of the groups and individuals who have taken up their democratic right.

CHAPTER 7
TOWARDS A JUST ETHIC OF PEOPLE, PLACE AND PROFIT

The systematic study of a conflict from the vantage points of its three key stakeholder groups – the local community, the state government and Alcoa – has thrown up a raft of social, place-based and environmental justice issues which together describe the unsustainability of a rural town and its community. We believe that this unsustainability is primarily the result of irreconcilable interests between local people and their corporate neighbour and elected leaders.

This unsustainability we also consider to be the result of systemic failings on the part of the company, government and society at large. We believe it to be born out of flaws within the assumptions underlying the corporate practices and government conduct this book has sought to describe. These failings and flaws are the subject of this chapter, which draws attention to the social and environmental trade-offs associated with the pursuit of a narrowly defined form of development. We attempt also to distil the lessons that can be learned from the Wagerup experience and identify pathways for future conflict avoidance.

The story told in this book, however, is also a case study of hope. Despite the trauma caused by the Wagerup controversy and the long list of failures that served to intensify it, the conflict has also galvanised members of the local populace who have acted with great courage and tenacity through the years of acrimony. We wish to highlight the bravery of some of the local heroes whose dogged pursuit of natural justice has met with some success.

In the last chapter, we offered a detailed analysis of the unsustainability of the Wagerup situation, the winners and losers of the conflict and the prevailing power dynamics. Still to be addressed are the desirability of the directions taken at Wagerup and the question of what should be done about it [1, p. 145]. These questions are heavily value-laden and speak to the deeply political nature of the controversy, which first and foremost ought to be seen as a power struggle over ideas, rights, interests and what counts as acceptable trade-offs and outcomes. We have seen throughout this book that the power struggle was uneven – it produced few winners and many losers. The mechanisms of power at work at Wagerup set out below are not mutually exclusive. Indeed, their interplay had the compounding effect of affording Alcoa sustained control over key determinants of the conflict.

MECHANISMS OF POWER AT WAGERUP

- Alcoa enjoyed legal protection to operate and expand at what was recognised as an unsuitable site, and continued occupancy of the territory – including the formally unsanctioned purchasing of private properties – permitting an even larger footprint for the refinery.
- Alcoa had substantial access to human resources and research to maintain the credibility of rational, technical knowledge at the expense of lived, emotional and local expert knowledge.
- Alcoa had control over its own public statements and corporate image and was thus able to prevent internal company differences from affecting its combative and defensive media strategies.
- Alcoa appeared to have allowed slippages and doublespeak between local, national and international promotional material, media releases and statements

to government inquiries. It maintained a stance of persistent denial of harm from pollution and adverse socioeconomic impacts at its Wagerup refinery.

- There are perceptions of collusion between Alcoa and the state government, particularly with regard to the hands-off approach by government and the seeming ineffectiveness of the state's environmental and health watchdogs, such that Alcoa was able to back away from considerations of compensation relating to the land management issue when the government promised to implement the Yarloop sustainability package as part of a regional planning strategy.

- Alcoa acted in contradictory ways in the Yarloop area with statements that it wanted people to stay in the area and for Yarloop to share its vision of a prosperous future, while undermining the very bases of social capital and community viability. Alcoa proved instrumental in the dismantling of a vibrant and cohesive community despite corporate communications claiming a central role for the company in the region's sustainability.

- Ameliorative efforts by Alcoa and the government were perceived as still advantaging Alcoa to a greater extent than Yarloop residents.

- To date, all forums which intersect with aspects of the controversy have attended to Alcoa's operational and expansion interests. There has not been the same dedication of labour and expertise from Alcoa to the community to serve its interests and visions for the future, where these interests and visions might not be acceptable to Alcoa.

- Media releases enabled concerned locals to get their stories out to a wider audience and were perceived to be the main reason Alcoa came briefly to the dialogue table. However, the media reports served to identify

'the complainers' who were seen as troublemakers by Alcoa and were treated unfairly as a result. Many of those who raised concerns and complaints found they were consequently disparaged and not treated as valued stakeholders.
- The extent of community outrage [2] was acknowledged in Alcoa's Donoghue and Cullen's research [3], yet company outrage, which appeared to underpin Alcoa's defensive reactions at key points in the controversy, has yet to be acknowledged. Alcoa could keep its emotionality behind closed doors and controlled public statements, maintaining a façade of the factual and rational compared to what was portrayed as an emotionally charged, irrational community.
- The rural location, small population and largely unseen impacts of Alcoa's actions away from the city, the seat of government and the majority of voters, meant that getting the issue onto the political agenda and getting effective action was very difficult.
- There were perceptions that the ECU research in 2002 and 2003 was compromised by Alcoa ceasing to authorise the community-based meetings and problem-solving efforts [4].

There can be little confidence that the dynamics of the Wagerup controversy have changed. The power asymmetries resulted in an imbalance between local costs and corporate gains, leading to an outcome we consider unsustainable and unjust. Injustices have arisen from corporate and political transgressions of what we described as 'warrants' in Chapter 6. These warrants speak to the ethics of company, community and government conduct and challenge all parties to cultivate new 'tactics and strategies' to achieve mutually beneficial and desirable processes and outcomes. Importantly, these warrants entail an order of accountability for each group, which always and ultimately is to the most vulnerable: the voiceless [e.g., ecosystems, see 5], the silenced [6], the most outspoken, and the pathologised absent [7]. According to social justice imperatives, accountability must be to those most at risk of being marginalised and made bad, wrong or invisible as part of any power struggle [8].

The .... form [of ethical sensibility required] suggests the necessity and potential of a dialogue across the corporate boundary with those most vulnerable to the effects of corporate conduct [Hamann citing 9, p. 358].

In Yarloop the most vulnerable were the losers of the long-running conflict, whose interests those in the privileged positions of power failed to protect. In fact, their vulnerability may have enhanced corporate control of the conflict parameters. That is why we propose here a socially just, compassionate and dialogical approach to the resolution of industry–community conflicts as a way of ensuring that economic development is not only profitable but is actually sustainable.

IS IT DESIRABLE?

The situation at Wagerup is not desirable according to any of the key stakeholders' self-confessed beliefs and values. There are transgressions against the state government's reckoning of what constitutes sustainable regional development [10], Alcoa's own mission and values statement [11] and the social and environmental values of people in the town and district of Yarloop. While on the face of it, all groups operated within compatible sustainability frameworks, a dominant, narrowly defined economic understanding of sustainable development prevailed.

This dominant understanding, which is mirrored nationally and internationally, favours a development path that promises employment, income, investment and tax revenue at what are considered to be acceptable social and environmental costs. These costs go largely unrecognised in the marketplace and are not captured or measured on corporate balance sheets or in national accounts, with the result that the benefits of the enterprise are overstated. Moreover, those bearing the costs of this form of development are often found to be voiceless in the negotiation of the dominant social contract, which envisages growth without limits. As the case study data showed, the financial gains of the company and the state came at considerable socio-ecological and psycho-social costs born by members of the local community – loss of sense of place, community cohesion and social connection. These asymmetries and their rarely questioned social acceptability lie at the heart of the growing critiques voiced on globalisation and economic growth. Their qualitative, almost anecdotal nature often defies quantitative measurement, which is why they frequently remain part of the hidden costs of development [12].

Wagerup can be understood as a microcosm reflecting what is a global problem: balancing the costs and benefits of economic expansion. The dominance of the conventional pro-growth ideology, supported by the kind of econo-political entanglements we detected in our analysis of the Wagerup controversy, leaves no space for a critical engagement with the pros and cons of development. Members of the Yarloop community and surrounding areas learned first-hand how, in the face of considerable local evidence of undue harm and impact, an apparently 'united front' of industry and government was able to maintain a business-as-usual approach at Wagerup. Local demands for fairness and justice were countered with economic and scientific arguments of acceptable risk and the greater good. Questions of acceptability were determined by those safely distant from the risks community members were exposed to. The undemocratic, utilitarian justifications for the acceptability of risk represent a prima facie case of injustice as those exposed to risk did not participate in negotiations about what was considered acceptable [13]. Compounding the injustice, these decisions were based on imperfect, vested science with, as time has shown, considerable knowledge gaps, leaving the community in the error margins of biased corporate and political decision-making.

The mechanisms of power we identified earlier ensured that the status quo of dominant interests, while challenged by a vocal minority, remained unchanged. Dynamics such as these reflect abuses of power by way of oppression and injustice. The forms of oppression visible in the Wagerup controversy can be summarised under what has been termed the five faces of oppression [14]:

- *Marginalisation* – the centre–periphery dynamic which resulted in predictable groups being sidelined and their needs and rights minimised.
- *Violence* – Alcoa's divide and conquer strategies, for example, as reported in individual land management negotiations.
- *Cultural imperialism* – Alcoa's control of language and definitions of what counts as rational knowledge and as a legitimate issues of debate (e.g., odour, dust excursions, etc.).
- *Powerlessness* – the failure to protect the rights of residents in the face of a corporate and government pro-growth agenda.
- *Exploitation* – Alcoa's free use of residents' time and knowledge in hundreds of meetings and conversations over many years.

A family protests against Alcoa (Photo by H. Seiver)

The centre-periphery aspect of the conflict and the associated power asymmetries gave rise to what is described in the literature as 'placeless power' and 'powerless places' [15]. Yarloop became a 'powerless place' while power resided in the senior boardrooms of Alcoa and the WA government during the years of conflict. Even though its Wagerup refinery was at the centre of the conflict, Alcoa could exercise placeless power. It could also argue the business case for its 'excellent' sustainability efforts by conflating and generalising its survey results of other neighbouring communities' satisfaction with Alcoa throughout Australia. This had the effect of diluting the seriousness of the results from the Wagerup survey and disempowering local experiences and places. As a US-based multinational corporation with executive managers able to influence decisions of governments across borders, Alcoa exercised placeless power while at the same time maintaining a 'powerful place' at Wagerup by occupying the territory and pursuing its commercial interests. It is this aspect of 'placeless power' or 'remote control' that forms the basis of the pending class action against Alcoa in the US court system, for it is argued by the claimants that decisions on the Wagerup operation were made in Pittsburgh as opposed to Perth or Wagerup [16, Section 17].

The exercise of 'placeless power', as the case study data showed, imposed high costs on a 'powerless place' with the continued and expanding operations of Alcoa at Wagerup. Further, while existing losses remain uncompensated to the extent residents believe would be fair, the perceived threat of a future expansion of the refinery suggests that the burden on the community is not yet contained or finished. The power dynamics remain unchanged.

FLAWED THEORY ... FLAWED PRACTICE

We introduced Alcoa as a company that sees corporate social responsibility as an integral part of its genetic make-up. Alcoa subscribes to what we referred to earlier as dominant, capitalist strands of corporate social responsibility (CSR) theory, which are premised on an assumed compatibility between a company's profit motive and broader social and environmental imperatives [17; 18]. Yet, in light of the Wagerup experience one ought to question the extent to which the pursuit of corporate self-interest can help address those social and environmental concerns which lie beyond the business case for CSR [19; 20]. Indeed, the Wagerup case made explicit that such assumptions are convenient for those favoured by the econo-political status quo, for they ignore the power imbalances at work between multinational companies and small country towns and their communities. Alcoa's CSR agenda was geared towards the improvement of the company's reputation, its financial situation and competitive advantage, focusing chiefly on economic tangibles such as employment and income generation as well as community investments and philanthropic spending. Corporate communications cited in this book show clearly the economic framing of Alcoa's social responsibilities, which the company linked to its level of alumina production.

Those at the receiving end of Alcoa's CSR practices had no input in defining the company's agenda and strategies. The company was thus not likely to arrive at CSR outcomes they could trust to be effective or reliable, as its CSR agenda was not inclusive of the views of the community members it was meant to serve – people with high personal stakes and relevant knowledge about local issues. Instead, local issues and the existence of problems at the community level were denied or discounted. From the company's perspective the issues have been resolved, as Alcoa's most recent sustainability report [21] makes no reference to the ongoing conflict at Wagerup.

On the question of CSR effectiveness, Alcoa and members of the Yarloop community reach diametrically opposed conclusions. While from Alcoa's point of view the situation improved over time in light of reduced complaints figures, residents saw their lives and their community changed dramatically. Alcoa speaks of a 'bright' and 'sustainable' future for Yarloop whereas the lived experience of residents points to something very different.

Proponents of dominant CSR theory continue to invest faith in the social efficacy of the business case for CSR, believing that corporate self-interest is to the benefit of companies and their host communities. But we need to question whether corporations can be trusted or expected to be socially responsible, for they represent nothing more than a legally binding contract between investors and company management, which has at its core the generation of profits for the benefit of stockholders [22]. Nobel prize-winning economist Milton Friedman said, 'A corporation is the property of its stockholders. And its interests are the interests of its stockholders' [23]. The economic underpinnings of the corporate venture do not bode well for the protection of those whose interests and rights align poorly with the corporate raison d'être of economic return. The Wagerup case made plain that Alcoa responded to community outrage in an economically and scientifically rational way. While this approach was consistent with the economic objectives of the firm, it proved incapable of attending to the scale and complexity of the problems experienced at the local level. The raft of non-economic impacts that were felt deeply among the Yarloop community was reduced to a matter of land management and financial compensation. We therefore question both the ability and willingness of body corporates to operate beyond the parameters of economic efficiency and to meaningfully address people and place-based issues, especially those that do not make but cost money.

The ethical dilemmas multinational companies are facing often fall outside the clearcut boundaries of cost-benefit analysis [24; 25]. These are problems for which there are no binding legal or economic imperatives or clear rules to follow. Economists like Friedman regard corporations as amoral [23] in their disputing the existence of social responsibilities of the firm outside the marketplace and the letter of the law. Consequently, these responsibilities fall onto individual managers who find themselves in the invidious position of needing to make moral judgements and to balance the at times conflicting demands of different stakeholders. Given the complexity of the task, they need to be pragmatic and strategic in determining the legitimacy of stakeholders and CSR issues [26]. In the end, decisions such as these are based on managerial discretion and values, which often favour so-called 'definitive' stakeholders (shareholders, customers, employees) but tend to exclude 'dependent' stakeholders who have legitimacy and urgency but limited power [27; 28]. Just so, the legitimate claims of dependent stakeholders – members of the Yarloop community – were ignored or discredited by Alcoa, making the Wagerup experience an exemplar case of a stakeholder management approach that lacked inclusiveness and failed to account for power differences and risks to the community [29].

A number of individual decisions proved decisive in the Wagerup conflict and highlight the relative invisibility of dependent stakeholders, whose interests could be effectively overlooked, ignored or actively sidelined. The list below, which is by no means exhaustive, also shows that after all it is the decisions and actions of individuals that determine corporate conduct and the public face of the company.

FATEFUL ACTIONS AND DECISIONS AT WAGERUP

- Failure to communicate and act upon the emission data from the liquor burner at the Kwinana refinery even though human health impacts were highly likely.
- Denial of problems at Wagerup despite considerable health complaints among staff and residents.
- Applying for expansion of the Wagerup refinery despite repeated promises to the community that it would not seek to expand the operation at a time of heightened community concern about refinery safety.
- Undervaluing of local properties during property purchase and differential treatment of people within and between land management areas A and B.
- The mocking of Yarloop residents by company staff about their refinery-induced illnesses.
- Continuing land management meetings with the

community, thus keeping expectations raised that the collaboration to achieve fair treatment for property owners was supported by senior management.
- Alcoa's reneging on a key feature of the agreement reached with the community.
- Failure by senior Alcoa management to meet face-to-face with the community and engage in robust dialogue to achieve just outcomes.

Previous chapters provide rich detail on how the decisions and behaviours of Alcoa staff affected the dynamics between the community and the company and played a critical role during the years of conflict. Problems associated with the discretionary decision-making affected community health and wellbeing. Discretion was exercised in what proved to be ethical grey areas for a company that boasts considerable expertise in the field of engineering and matters technical but not in the area of managing complex social issues. It is thus our contention that in the absence of legislative guidance, Alcoa's CSR approach was ill-equipped to provide a reliable moral compass for dealing effectively with community grievances. The ability of individuals to make ethical and compassionate decisions was compromised by an in-house logic that, in the rational pursuit of the business case for CSR, failed to account for local issues that fell outside it. Those staff members who could relate to the local context reportedly found themselves replaced. In this way, Alcoa was able to maintain an inward-looking CSR approach that proved resilient to challenges from both inside and outside the company.

Alcoa operated without external checks and balances; it defined the agenda and determined its own measures of success. Fortunately there is a growing recognition that companies should not be left in charge of setting the CSR agenda, reflected in calls for new laws and social contracts that spell out social and environmental criteria for companies' licences to operate [30; 31].

DEVELOPMENT – A BLINKERED IDEOLOGY

The development maxim adhered to by successive state governments in Western Australia has placed at risk the sustainability of a small community, which due to no fault of its own came too close to one of the engines of the state's economy. The government's economic credo, coupled with the relative electoral insignificance of the Yarloop area, effectively overrode local concerns about the refinery. Yet the approach by government was consistent with the neo-liberal, economic rationalist stance which 'accepts and advocates the primacy of the markets' [32, p. 3] and believes in small government and big business. Indeed, the state government, itself a beneficiary of economic growth, appears as an almost invisible stakeholder in the conflict. Ministerial decisions and departmental responses to refinery related matters only served to fuel local perceptions of collusion, which were echoed by political insiders.

In today's corporate society, governments are frequently criticised by industry interests for being in the way of growth and development, too heavy-handed in their regulation of industry and bogged down with bureaucratic red-tape [33]. Indeed, recent years in Western Australia have witnessed many calls by industry for government to fast-track development approvals [34]. In Wagerup however, the government proved instrumental in the establishment and growth of the industry. Alcoa's operations enjoyed support and protection from the highest political office in the state from the day of their inception. The data presented in this book are indicative of the zeal with which development goals were pursued by the state, making large concessions to the industry concerning access to natural resources and its requirements for infrastructure, energy and water. These access rights are protected to this day through a state agreement which offers disproportionate industry protection and a possibly unprecedented publicly funded industry subsidy. Heavy-handed regulation and undue government interference would threaten industry investment in the state and jeopardise the growth project. The community was seen as a threat to foreign investment and therefore reprimanded by then Premier Alan Carpenter for being an inconvenience to both industry and government [35].

This lament is not new. State governments in the past frequently stood accused of being overly pro-development [36; 37] and too closely aligned with industry interests [38]. In Western Australia in particular, a 'soft touch' by government on the resource sector is detectable [39], evidenced by a string of decisions (or lack thereof) that advance the state's development agenda despite high-level recognition of the drawbacks of this approach. The state's environmental watchdog warns that WA's development trajectory, which is chiefly based on the large-scale exploitation of natural assets, is unsustainable, placing at risk environmental systems and services and affecting human wellbeing [40; 41]. But while the pro-development stance has long been subject to criticism, development continues at an increasing rate, with the state's largest resource development project (the Gorgon gas project) recently approved by government [42].

The mindset underlying this political approach to development bears the hallmarks of a blinkered economic rationality [43], which dominates the political debate in Western Australia as well as in other states and at the federal level [44]. It is a rationality that often goes unquestioned because its objectivity and neutrality are asserted strongly [45; 46] while its values and assumptions are hidden [47; 48]. This allows for the legitimising of economically rational policy prescriptions and the easy dismissal of non-economic perspectives, framed as value-laden, ideological and irrational [49; 50]. In extreme cases, as occurred federally during the years of the Howard government in areas such as climate change, media independence and the arts, dissent is actively silenced [51].

Supporters of economic rationalism's rejection of dissenting viewpoints is highly problematic because of the poor track record of economics in dealing with messy social and ecological problems [38]. Wagerup was a messy case, requiring a holistic and integrative understanding of interrelated problems that were not limited to economic aspects. It required the close cooperation among government and its key departments of health, environment, industry and resources for collective learning [52] to occur and to arrive at integrative solutions [53]. The proposed CSIRO study mentioned earlier, which was rejected by both government and industry, would have been such an approach, for it was an attempt to gain a holistic understanding of the community impacts of the Wagerup facility. But, the economically rational lens of government meant that there was little support for the in-depth exploration of issues and themes that industry and government gave little credence.

In this book we have questioned the approach taken by government and its underlying assumptions in light of the social and environmental fallout they incurred. Even from an economic point of view, the figures appear highly suspect [54]. The many perspectives retold here certainly suggest that the benefits of the government's development agenda were to a large degree negated by the cost to the community and the environment. Even if the economics are sound, a case yet to be made, questions of justice and fairness remain as they relate to the treatment and compensation of affected community members.

A COMPLICITY OF SILENCE

Earlier chapters spoke of the media coverage the Wagerup conflict received at the local, national and international level during the height of the controversy between 2001 and 2003. Yet despite the publicity, the case has not attracted the attention of the social mainstream or resulted in widespread public demands for government intervention. Somehow Yarloop did not make it onto the public radar for a sufficient time. The debate thus largely occurred outside the gaze and attention of the general public.

The silence of the public is also an expression of the centre–periphery dynamic, illustrating the difficulty of getting localised rural issues recognised in places of electoral significance in urban Australia. By contrast, the lead poisoning incident in Esperance in 2007, a popular holiday destination for many Perth residents, caused much public indignation in response to vivid pictures in newspapers and on television screens of thousands of sea birds dying. The incident was followed by a relatively prompt and far-reaching political response [55; 56; 57]. Yarloop never reached this iconic status.

The silence of the wider public is of concern for a number of reasons. First, silence provides corporations with a social licence to operate and legitimates corporate practices. Secondly, silence works against political and legislative change. Corporate law, just like any other local, state or national law, determines rules of conduct and as such is a reflection of a society's values and beliefs. Silence can thus be assumed to be an expression of a social consensus that laws by government and corporate conduct are in line with what society believes to be good and proper [22]. We challenge this assumption in the belief that the

Wagerup experience would violate anyone's sense of justice and fairness. The lack of a stronger public reaction to media reports about the state of affairs in Wagerup served to isolate the conflict and deprive residents of the critical mass needed for political mobilisation. The legitimacy of corporate conduct which was provided by experts, government officials and Alcoa's own interest groups was given tacit assent through the silence and conformist stance of the public mainstream. Injustice can breed a 'culture of silence' which further advantages the power elites and creates a pervading sense of inevitability of the way things are [6]. This has been referred to as the 'unconscious civilisation' [58].

Policy making is often reactive, to new scientific insights or in response to an accident. Industry regulation in Western Australia was shown to lack the sophistication of the industry it was meant to govern. In the absence of a large industrial incident (such as occurred at Esperance) or scientific proof of causality, residents were the only living proof of there being a problem.

WHAT SHOULD BE DONE?

The question of 'what should be done' [1] is a matter of perspective. We declared our hand early, arguing in favour of a social change agenda and on the basis of a considerable body of literature that spells out principles of justice, fairness and sustainability. Our views are informed by our own research experience as well as the growing critique on the econo-political status quo, which in our assessment underlies the conflict this book describes. We do not wish to prescribe new rules for industry–community engagements and policy making by government. Instead we offer our understanding of the conflict and provide a set of critical prompts that may inform a needed dialogue between communities and their commercial and political leaders.

Compassionate use of power

We regard an awareness of and critical reflection on corporate power as the essential basis for stakeholder engagement. The micro practices of power in everyday interactions [59] as well as the extra-local relations of ruling [60] help explain the importance of such awareness and reflection. At the micro level (referring here to little, everyday interactions), power and knowledge are enacted in particular relationships and situations. According to the case study, the corporate enacting of power and knowledge had the cumulative effect of disadvantaging Yarloop and advantaging Alcoa and the government. At the extra-local level (beyond the local level) of power relationships, Alcoa as a corporate entity had pre-set rules, procedures and business assumptions which drove its key decisions. Alcoa's institutional practices were not geared towards deference to aggrieved community stakeholders. Thus, according to their institutionalised logic, Alcoa staff could not act in the interest of the community at the expense of their company's profit and image-related motivations.

Logic, however, is not the only basis for making decisions, especially where such disparity of power and irreconcilable differences are involved. Alongside the many stories of offence, harm and loss recounted by Yarloop people at the hands of Alcoa, ECU researchers witnessed a range of genuine efforts by individual Alcoans to understand the issues from the community's perspective(s).

At times during land management meetings, exchanges had elements of respect, care, relevant knowledge and attempts to respond fairly to the issues. This is what is described as the exercise of power based on a love ethic [61; 62]. A love ethic suggests it is incumbent on powerful parties to ensure they are not dominating or oppressing their marginalised, vulnerable or even resistant stakeholders [8]. Alcoa's stated commitment to values-based business practices [11] is commendable, but it has fallen short of local expectations. Ethics and values unhinged from a critical awareness of Alcoa's own power abuses [63] open the way for the type of corporate power experienced at Yarloop.

Alcoa is not a monolithic, fixed entity [64] with a predetermined blueprint that can be perfectly imposed in isolation from the legal and socioeconomic context in which it operates. We have uncovered inconsistencies,

Acknowledging Alcoa's help (Photo by H. Seiver)

contradictions and differing points of view within Alcoa with regards to what went wrong and what was needed to make things right. There were complex interplays of loyalties, access to information and ability to influence the agenda. For example, one of the most challenging positions involved Alcoa employees who were also residents of Yarloop, with family members at odds with their own perceptions of the situation. Some of these people were also key community leaders trying to get Alcoa and the government to be accountable for the problems. There is at least one instance of a person who was initially pro-Alcoa and did not believe their partner's illness was due to refinery emissions, who came to be highly critical of the company.

For otherwise loyal employees to become activists against their employer and thereby risk their jobs, indicates something of the scope for resistance to centralist control by Alcoa. That some of the most influential critics were also employees of Alcoa, gives some hope that Alcoa's controlled public story might give way to a more reflective, compassionate and non-combative positioning, which recognises the wisdom of the people [65]. If so, there might be some chance of Alcoa returning to the dialogue table more willing to be equal partners – and a glimmer of hope for a more sustainable future scenario for Yarloop.

We might all agree that communities have the right not to be harmed by industry [66]. Yet, the onus of proof often rests with affected individuals, whose non-expert status prevents them from having their claims accepted as legitimate knowledge. Alcoa's scientific determinism rejected local knowledge and prevented the company from engaging more openly and critically with the issues at the heart of the conflict.

We advocate shifting the onus of proof onto companies – especially in industries where much scientific uncertainty about impacts still remains – along with a stronger push for scientific competency in decision-making processes on the policing, governance, and control of potentially harmful industries [67]. It is also important to recognise the value of non-experts for their local knowledge and venerable experience, because their insights are known to 'add value' to the work done by so-called experts [68; 69; 70; 71; 72; 73]. As an external consultant remarked:

> ... *[the community is] sitting in a dispersed network in proximity to this facility and if a plume strike is coming to ground at one point, that's a complaint but it's from a detailed antenna, a dense antenna, far denser than the department or Alcoa themselves could actually set up and install and operate. They hear it but they don't necessarily weigh it as validly as their own direct measurement.*

The marriage of expert and non-expert knowledge requires a preparedness by the gatekeepers of knowledge to change acculturated assumptions about 'what is' and to act upon new insights should they point to the existence of a different reality or scenario [74].

Despite the enlightened rhetoric about the social responsibilities of the company, for many companies the business of business remains business [75], and there are limits to the level of trust that can reasonably be invested in the conduct of profit-seeking entities. We have spelled out our desire and rationale for compassionate, dialogical approaches to stakeholder engagement by companies. Yet we are also aware that competitive pressures and stakeholder demands, as well as perverse market signals and cost arguments may lure companies onto the thin ice of discretionary decision-making that is prone to leaving vulnerable company stakeholders short-changed. While corporations such as Alcoa claim to be guided by the 'new rationale' of 'enlightened value maximisation' [76; 77], social, altruistic and humanitarian interests seem to remain outside the corporate, economic mindset [24; 78] and continue to fall outside the 'firm's proper scope of activities' [25, p. 605]. Under the banner of CSR, companies may seek to socialise and legitimate their operations [79], yet often fail to challenge their fundamental values [80]. What remains is business-as-usual with a social twist.

The prevailing, and we believe growing, disconnect between body corporates and local communities is well documented [22; 58; 81; 82; 83], and terms such as 'corporate social responsibility' and 'business ethics' continue to be seen as paradoxical and oxymoronic [84; 85; 86]. Despite noteworthy but isolated evidence to the contrary, we regard assumptions about the coinciding and overlapping of corporate interests and those of society as naive at best and cynical at worst.

As the case study and our analysis have shown, complex social and environmental problems demand a more hands-on approach by the regulator as well as an active citizenry to hold corporate decision-makers accountable and to minimise or mitigate harm. The community in this story has been fighting vehemently to 'reassert the local' [87] and to claim back power. These efforts, however, were not matched by the general public, and nor did the regulator assume responsibility and sufficiently assert the rights of the constituents it is meant to represent and protect.

Deliberative, participatory democracy

We suggested earlier that Western Australia has a preference for a small-government, big-business approach with an emphasis on industry self-regulation and minimal government interference [88; 89]. Most measures taken by government rely on strategies which are generally non-binding statements of government intent, focusing largely on partnership-building between industry and government, voluntary action and cooperation [90]. Little is being done by way of direct regulation, which also explains why state legislation in many areas is found to be weak and ineffective as well as often short-lived or poorly implemented. Primacy continues to be given to economic over social and environmental concerns [88; 91]. Despite being the first state with a formal sustainability strategy [10], Western Australia continues to follow a weak sustainability policy path at best [92].

We pointed to the ideological constraints at the heart of the government's stance on industry regulation – the failure to protect those living downwind of companies favoured by government's strong faith in unfettered markets. This is compounded further by the unwillingness of the regulator and its departments to engage with the electorate on social and environmental matters that have direct implications for business interests. This form of governance is resistant to stakeholder input and external changes, as problems are framed, discussed and decided upon by political and corporate elites. Public participation and scrutiny are not encouraged while information remains tightly controlled. This results in a highly constrained political debate, serving only the convenience of the political system. Political decisions are made ad hoc and under pressure, creating what has been coined 'garbage-can' policies [93] such as the government's supplementary property purchase program (SPPP), which was widely criticised for its poor design and implementation.

By contrast, governance for sustainability is based on empowerment, the dispersal of power and responsibility, a shift away from passive compliance to the sharing of governance [94]. This form of governance is responsive to community input and external change, and thus has active-adaptive capacities that enable robust and effective policy making, which becomes proactive and self-corrective. Applied to the Wagerup experience such an approach would result in an early political response to signs of local distress and ill-health. Steps would be taken to ensure that complaints are investigated and that company health and emissions data are verified and tested independently. Overall, commensurate with the harm potential of the industry, regulation would be devised on the basis of robust, independent scientific advice and adjusted in response to new or emerging evidence which would include local experience. Questions pertaining to acceptable risks would be determined jointly with the communities likely to be affected, and the negotiation of pathways for harm mitigation or, if necessary, compensation would involve all affected parties.

Trust is another critical pre-condition for good governance [95]. Participants in political processes 'have to enter on the basis of trust in the processes in which they are operating, and in the spirit of reaching an outcome that is fairly and respectfully arrived at' [94, p. 100]. Trust gives legitimacy to political processes. However, trust has to be actively produced and negotiated [96] over time. Trust is developed from a background of 'trust culture' and dependent on a structural context (e.g., inclusiveness and fairness of government processes) and so-called agential endowment (e.g., social moods and collective capital) [97].

During the Wagerup conflict little scope existed for the development of trust between the Yarloop community and the government because of both the prevailing structural context and agential endowment of the negotiating parties. The closedness and inaccessibility of government and it departments severely limited the community's ability to influence the political process and decisions. Trust was also hampered by a string of government decisions which only served to further alienate Yarloop residents.

'[A]t the end of the day it is a matter of trust,' residents were often heard to say when commenting on the nature of the Wagerup conflict – but there was no trust, neither with the government nor with Alcoa.

The community's lack of faith in the fairness of government decision-making processes that seemed biased towards industry only served to heighten levels of community distrust in government. Active steps are needed to repair fractured relationships, starting with a genuine engagement with the concerns of residents.

The implementation of good governance principles is chiefly dependent on the willingness of decision-makers to relinquish direct power and to engage in open dialogue with their constituencies. With such a transfer of power comes the transfer of responsibilities. The most fundamental civic good lies in the willingness of the public to participate in political processes on matters of public interest. Powerless places like Yarloop require active empowerment to provide a platform on which negotiating parties can meet as equals. Western Australia's political and bureaucratic structures, however, with their closeness to commercial interests and neo-liberal, pro-growth ideology, appear to be key obstacles to enablement of such deliberative processes. Just as industry cannot be relied upon to put people ahead of profits, governments with strong pro-growth convictions are equally prone to put development goals ahead of social and environmental imperatives [98].

The dogged pursuit of justice

A well-informed active citizenry is crucial for the sustainable and just resolution of conflicts [99]. The Wagerup experience illustrates how corporate dominance can pose a threat to the workings of the democratic system. The quest for profits and competitive advantage has meant that over the years corporate influence in political decision-making processes has grown, while the capacity of ordinary citizens to be heard and have agency in the political realm has diminished.

Concerned and informed residents have been active in Yarloop, forming groups that have evolved into extensive networks with access to a wide range of expertise. The Community Alliance for Positive Solutions (CAPS) boasts a local, national and international network of technical and legal experts, sympathisers, supporters and fellow community activists [100].

Despite the absence of overt support from the wider populace, our research allowed us to witness striking examples of an empowered and informed citizenry, active in the pursuit of justice for people and place. Local activists made a difference to the outcomes at Wagerup and lent their strength to efforts elsewhere through alliances with similarly affected communities and the creation of a web of active citizens around the world. This strategy of linking with other citizens is made more potent in its capacity to impact policy processes through the effective use of the internet and other communication technologies [100].

Local activists brought the situation at Wagerup to the public's attention, pressed the government to act and held Alcoa accountable for breaches of its operating licence and the impacts of its operations at Wagerup. These activists may have been unable to stop the dismantling of their town, but they have been successful in effecting changes in the conduct of Alcoa and government departments, and bringing about a series of far-reaching improvements in the way in which the industry is monitored. Without their efforts, the health concerns and land management impacts would have stayed hidden and been construed as 'private troubles' when in fact they constituted matters of public concern [101]. Their dogged pursuit of justice is testimony to democracy, as imperfect as it is, at work. These groups have kept the Wagerup issue alive. As one scientist remarked:

> ... Through the community poking its nose in and saying to our politicians that this is unacceptable, we need tighter regulations on this industry ... we've already seen a significant improvement in the government resources required to monitor this industry as well as the additional monitoring requirements then placed on this industry by the government in order to bring their environmental standards up.

Active citizens protest (Photo by H. Seiver)

CAPS remains the only stakeholder group that has offered a solution to the conflict. The idea to relocate Yarloop away from the refinery was raised in 2002 in a meeting with Alcoa management. It was dismissed as fanciful and 'fell off the table' as a serious option, but the idea is starting to gain political traction. Since 2009, the proposal has been considered by state and local planning authorities as a possible option, and CAPS is able to have input into the planning process. As one public servant remarked, it was the community who continued to offer win-win solutions as opposed to the government or the industry.

> *... Even after all of the effort ... and even after all of the community support and alliances and after everything that's gone on, the community still stands strong and they still are able to find potential avenues that our government should be taking and this industry should be taking, because of all of the discussions that I've had with people, it's not that they want to shut Alcoa down, but they want to see Alcoa have a better management practice and what they want to see is getting their lives back. They're not ... opposed to saying why can't we have a win-win situation? Why can't we get the best out of this industry and get the best out of our community, and here are some possible areas that we could target.*

Members of the various community groups have given thousands of hours of unpaid work in meetings, writing submissions, running workshops – and the myriad other tasks that make up such an extended campaign – and most of it in the face of staunch opposition by government, industry and parts of the community. At last they are getting some late recognition. In the last few years, members of CAPS have been invited to participate in discussions about licensing conditions for the refinery, regional planning schemes, questions of licence renewal and air quality monitoring and Alcoa's proposed recommissioning of its oxalate kiln [102; 103; 104]. CAPS also successfully negotiated a memorandum of understanding with the Department of Environment and Conservation, in which both parties agreed to full information sharing and to collaborate in their air monitoring efforts in the

Yarloop area [105]. CAPS members were urged by the Minister for the Environment, Donna Faragher, and the Department of Environment and Conservation to join the Wagerup Tripartite Group and to stipulate their terms for participation in the meetings between the department, Alcoa and the community [106; 107; 108; 109]. Members were also invited to maintain dialogue with the department outside the tripartite forum [110; 106].

In late 2009, representations were made by Alcoa and government officials with a view to involving CAPS members in negotiations [111]. It remains to be seen, however, whether this change of tack on the part of Alcoa and the government is motivated by the looming US court case, which could have serious implications for both the company and the regulator.

This dogged pursuit of justice has been a burden on activists. Many experienced mental and emotional burnout during the years of conflict, and for some the stress became overwhelming. A family member living outside of Yarloop said:

It doesn't matter when I go down there; every time I go down there we talk about Alcoa ... I just feel that that's affecting them health-wise, mentally, financially to the point that something's going to break. Something has got to give.

Indeed, many things have given way. Local heroism has come at great personal cost, in terms of time and finances, relationships and stress, and the associated mental burdens of what amounted, in some cases, to over ten years of campaigning. While local community activists expressed to us that they felt they were making a difference:

It's been hard yakka ... I believe it's been all worthwhile. And I think ... we have made a difference. And things

> *are still happening, which, hopefully, will get us to the point that we want this to go.*

... and that they were committed to seeing their campaign through:

> *I will see this through ... this is what it's all about ... to get some sort of accountability for the little people.*

... the costs were also being counted:

> *A lot of expense ... a lot of hours.*
>
> *A lot of cost ... personal cost.*
>
> *But everything is CAPS, CAPS, CAPS. Our whole world revolves around CAPS. All we talk about is CAPS, CAPS, CAPS. It works fine at the moment, but ... I am really sick of it ... I do want to continue to work for CAPS, but I am limiting how much time I am putting into it, just for my own self-preservation.*
>
> *I'm worried about what the stress does to me.*

WHY IT MATTERS

> *We need to learn as much as we can ... hopefully what we will see is a change in our government system to deal with these situations in the future ... the more that we learn and the more we can prevent it from happening in the future the better ... This isn't the only one; there are plenty of other examples out there where communities are fighting industries that are too close. (public servant)*

We stated at the outset that the story of Yarloop is mirrored in many places around the country, and indeed worldwide. We have sought to make explicit the plight of residents in an attempt to initiate a broader debate on the desirability of narrowly defined development trajectories

Alcoa the wrecker (Photo by H. Seiver)

and a reflection on the question of the greater good. We hope to have shown, based on the data presented in this book, that calls for a critical engagement with corporate and political pursuits of growth are well founded in light of the associated costs.

Australia's overt growth ambitions coupled with its rich endowment in natural assets and growing population, make it more than probable that the Wagerup experience will be relived in various parts of the country in years to come. We have catalogued some of the ingredients we perceive to be essential for the avoidance of future conflicts, integral to which is the knowledge and understanding of the impacts uneven forms of development and the uneven distribution of power can have. Equally important is the realisation, as demonstrated by the persistence of members of the Yarloop community, that changes within corporate and political spheres can be achieved for as long as communities continue to assert their rights and work towards just and fair outcomes.

Both government and the corporate sector have adopted the language of responsibility and sustainability under the banners of CSR and good governance. In the case of Wagerup, however, political and commercial interpretations were at odds with local understandings of sustainability and, indeed, were found to be undermining the very sustainability of the Yarloop community. The incompatibility between various perceptions of what sustainability entails demands debate on how these differences can be reconciled. Without such public engagement, we are unlikely to see much change in the air under (our!) corporate skies.

NOTES

INTRODUCTION

1. Ferguson, I., Lavalette, M., & Whitmore, E. (2005). Introduction. In I. Ferguson, M. Lavalette & E. Whitmore (Eds), *Globalisation, global justice and social work*. London: Routledge.
2. Saul, J. R. (2005). *The collapse of globalism and the reinvention of the world*. New York: The Overlook Press.
3. Whitmore, E., & Wilson, M. (2005). Popular resistance to global corporate rule: the role of social work (with a little help from Gramsci and Freire). In I. Ferguson, M. Lavalette & E. Whitmore (Eds), *Globalisation, global justice and social work* (pp. 189–206). London: Routledge.
4. Pell, D. J. (1996). The Local Management of Planet Earth: Towards a 'Major Shift' of Paradigm. *Sustainable Development, 4*, 138–48.
5. Australian Bureau of Statistics. (2006). *Australian national accounts. State accounts [Cat. No. 5220.0]* (No. 5220.0). Canberra: ABS.
6. Beard, J. S., Chapman, A. R., & Gioia, P. (2000). Species richness and endemism in the Western Australian flora. *Journal of Biogeography, 27*, 1257–68.
7. Sleeman Consulting, & Goodall and Business and Resource Management. (2004). *Energy for minerals development in south west coast region of Western Australia*. Perth: Western Australian Department of Industry and Resources.
8. Walker, K. J. (2001). Uncertainty, epistemic communities and public policy. In J. W. Handmer, T. W. Norton & S. R. Dovers (Eds.), *Ecology, Uncertainty and Policy. Managing Ecosystems for Sustainability* (pp. 262–90). Harlow, UK: Pearson Education Ltd.
9. Elkington, J. (1994). Towards the sustainable corporation: Win-win-win business strategies for sustainable development. California Management Review, 36 (2), 90–100.

CHAPTER 1

1. Australian Bureau of Statistics. (2002). *2001 census community profile series: Yarloop*. Canberra: ABS.
2. Alcoa. (2006). *Alcoa submission to the State Infrastructure Strategy*. Perth: Alcoa World Alumina.
3. Alcoa. (2007). Alcoa has been sustainably mining, refining and smelting in Australia since 1961. Retrieved 1st May, 2007, from http://www.alcoa.com/australia/en/alcoa_australia/australia_overview.asp
4. Corporate Knights Inc., & Innovest Strategic Value Advisors Inc. (2007). Global 100. Most sustainable corporations in the world. Retrieved 1st May, 2007, from http://www.global100.org/2007/index.asp
5. Reputex. (2003). *Social responsibility ratings 2003*. Melbourne: Reputex.

6. Standing Committee on Environment and Public Affairs. (2004). *Report of the Standing Committee on Environment and Public Affairs in relation to the Alcoa refinery at Wagerup inquiry*. Perth, Western Australia: Legislative Council.
7. Robertson, A. (2005). *Submission by the Department of Health - Wagerup refinery unit three expansion. Environmental review management programme (ERMP)*. Perth: Department of Health.
8. Wagerup Medical Practitioners' Forum. (2005). *Submission on ERMP: Wagerup refinery unit three expansion*. Perth: Wagerup Medical Practitioners' Forum.
9. McGowan, M. (2006). *Environmental approval for the Alcoa expansion. Statement by the Minister for the Environment*. Perth: Government of Western Australia.
10. Alcoa. (2006). Green light for alumina refinery expansion. Retrieved 1st May, 2007, from http://www.alcoa.com/australia/en/info_page/WAG_home.asp
11. Abramowitz, M. (1989). *Thinking about growth*. Cambridge: Cambridge University Press.
12. Australian Bureau of Statistics. (2006). *Measures of Australia's progress 2006* (No. 1370.0). Canberra: ABS.
13. Alcoa. (2007). Sustainability Approach. Retrieved 1st May, 2007, from http://www.alcoa.com/global/en/about_alcoa/sustainability/home_sustainability_approach.asp
14. Government of Western Australia. (2003). *Hope for the future: The Western Australian state sustainability strategy*. Perth: Department of the Premier and Cabinet.
15. Alcoa. (2005). *Your future our future*. Perth: Alcoa World Alumina.
16. Alcoa. (2006). *Summary response to appeals issues. Wagerup unit 3 expansion project*. Applecross: Alcoa World Alumina.
17. Carpenter, A. (2006). *Brief ministerial statement: Alcoa Wagerup expansion*. Perth: Office of the Appeals Convenor.
18. Environmental Protection Authority. (2006). Media Release - EPA Bulletin 1215 -Wagerup Alumina Refinery - Increase in Production to 4.7 Mtpa; and Wagerup Cogeneration Plant. Retrieved 1st May, 2007, from http://www.epa.wa.gov.au/article.asp?ID=2183&area=News&CID=18&Category=Media+Statements
19. Hamilton, C., & Saddler, H. (1997). *The Genuine Progress Indicator. A new index of changes in well-being in Australia* (No. 14). Canberra: The Australia Institute.
20. State of the Environment Council. (2006). *Australia state of the environment 2006*. Collingwood, Victoria: CSIRO Publishing.
21. Australian Aluminium Council. (2004). *Sustainability report 2004*. Canberra: AAC.
22. Alcoa. (2005). *Environmental review and management programme Wagerup refinery unit three* Perth: Alcoa World Alumina.
23. Office of Energy. (2003). *Energy Western Australia*. Perth: Government of Western Australia.
24. Schur, B. (1985). *Jarrah forest or bauxite dollars?: a critique of bauxite mine rehabilitation in the jarrah forests of southwestern Australia Perth (WA)*. Perth: Campaign to Save Native Forests (WA).

25. Western Australian Forest Alliance. (2007). Boycottt jarrah. Retrieved 5th May, 2007, from http://www.wafa.org.au/actions/boycottjarrah.html
26. Pell, D. J. (1996). The Local Management of Planet Earth: Towards a 'Major Shift' of Paradigm. *Sustainable Development, 4*, 138–48.
27. United Nations Division for Sustainable Development. (1992). *Agenda 21*. Rio de Janeiro: United Nations.
28. Adams, W. M. (2006). *The future of sustainability. Re-thinking environment and development in the twenty-first century*. Gland, Switzerland: IUCN.
29. Yarloop and Districts Concerned Residents' Group. (2005). *Submission against the Alcoa Wagerup refinery*.
30. Ross, D. (2007). When a rural town and global company meet – A community begins to find its voice in threatening times. Unpublished paper. Bunbury: Edith Cowan University.
31. Flint, J. (2006, 10th September). Alcoa expands despite toxins. *The Sunday Times*, p. 16.
32. Ross, D. (2003). *Reviewing the land management process: Some common ground at a point in the process. A report on the collaboration between Alcoa, Wagerup and Yarloop/Hamel property oweners*. Bunbury: Edith Cowan University.
33. Utting, K. (2002, 19th March). People we do have a problem – Alcoa. *Harvey Reporter*.
34. Basagio, A. D. (1995). Methods of defining sustainability. *Sustainable Development, 3*, 109–19.
35. Costanza, R., Mageau, M., Norton, B., & Patten, B. C. (1998). What is sustainability? In D. J. Rapport, R. Costanza, P. R. Epstein, C. Gaudet & R. Levins (Eds.), *Ecosystem Health* (pp. 231–9). Carlton (VIC): Blackwell Science, Inc.
36. Eckersley, R. (1998). *Measuring progress: is life getting better?* Collingwood, Victoria: CSIRO Publishing.
37. World Commission on Environment and Development. (1987). *Our Common Future*. Oxford: Oxford University Press.
38. Elkington, J. (1994). Towards the sustainable corporation: Win-win-win business strategies for sustainable development. *California Management Review, 36*(2), 90–100.
39. Giddings, B., Hopwood, B., & O'Brien, G. (2002). Environment, economy and society: fitting them together into sustainable development. *Sustainable Development, 10*, 187–96.
40. Lafferty, W. M., & Meadowcroft, J. (Eds.). (2000). *Implementing Sustainable Development. Strategies and Initiatives in High Consumption Societies*. Oxford: Oxford University Press.
41. Aplin, G. (2000). Environmental rationalism and beyond: toward a more just sharing of power and influence. *Australian Geographer, 31*(3), 273–87.
42. Dryzek, J. S. (1996). Foundations for environmental political economy: The search for homo ecologicus? *New Political Economy, 1*(1), 27–40.
43. Hamilton, C. (2002). Dualism and sustainability. *Ecological Economics, 42*(1-2), 89–99.

44. Hamilton, C., & Maddison, S. (2007). *Silencing dissent. How the Australian government is controlling public opinion and stifling debate*. Crows Nest, NSW: Allen & Unwin.
45. Handmer, J. W., Norton, T. W., & Dovers, S. R. (Eds.). (2001). *Ecology, uncertainty and policy. Managing ecosystems for sustainability*. Harlow, UK: Pearson Education Ltd.
46. Brueckner, M. (2007). The Western Australian Regional Forest Agreement: economic rationalism and the normalisation of political closure. *Australian Journal of Public Administration, 66*(2), 148–58.
47. Cuthill, M. (2002). Exploratory research: citizen participation, local government and sustainable development in Australia. *Sustainable Development, 10*, 79–89.
48. Davis, G. (1996). *Consultation, Public Participation and the Integration of Multiple Interests into Policy Making*. Paris: Organisation for Economic Co-operation and Development (OECD).
49. Innes, J. E., & Booher, D. E. (2004). Reframing public participation: strategies for the 21st century. *Planning Theory and Practice, 5*(4), 419–36.
50. Saul, J. R. (1997). *The unconscious civilisation*. Ringwood (VIC): Penguin Books Australia Ltd.
51. Theobald, R. (1997). *Reworking Success : New communities at the end of the millennium*. Gabriola Island, B.C.: New Society Publishers.
52. MMSD Project. (2002). *Breaking new ground*. London: Earthscan.
53. Jensen, M. C. (2002). Value maximization, stakeholder theory, and the corporate objective function. *Business Ethics Quarterly, 12*(2), 235–56.
54. Blowfield, M. (2005). Corporate social responsibility - the failing discipline and why it matters for international relations. *International Relations, 19*(2), 173–91.
55. United Nations Development Programme. (1997). *Reconceptualising Governance. Discussion Paper 2*. New York: UNDP.
56. Porter, M. E., & Kramer, M. R. (2006). Strategy and society: The link between competitive advantage and corporate social responsibility. HBR Spotlight. *Harvard Business Review, December*, 1–14.
57. Korhonen, J. (2002). The dominant economics paradigm and corporate social responsiblity. *Corporate Social Responsibility and Environmental Management, 9*(1), 66–79.
58. Ross, D. (2002). *Enacting my theory and practice of an ethic of love in social work education*. Bunbury: Edith Cowan University.

CHAPTER 2

1. Hughes, O. (1980). Bauxite mining and jarrah forests in Western Australia. In R. Scott (Ed.), *Interest groups and public policy: Case studies from the Australian states* (pp. 170–93). Melbourne: Macmillan.
2. Lines, W. J. (2006). *Patriots : defending Australia's natural heritage*. St. Lucia, Qld: University of Queensland Press. Alcoa. (2006).
3. The map is based on Alcoa. (2006). *Wagerup unit three - Overview of sustainability & permitting components*. Perth: Alcoa World Alumina.

4. Morton, A. (2009, 27th November). Secrecy on aluminium subsidies to remain *The Age*, from http://www.theage.com.au/environment/secrecy-on-aluminium-subsidies-to-remain-20091126-juon.html.
5. Turton, H. (2002). *The aluminium smelting industry: Structure, market power, subsidies and greenhouse gas emissions (Discussion Paper Number 44)*. Canberra: The Australia Institute.
6. Kelly, B. (1976, 16th September). Refinery site in Hamel area? *South Western Times*, p. 2.
7. Murray, P. (1976, 11th September). Big S.W. land deal mystery. *The West Australian*, p. 1.
8. Langley, G. (1976, 23rd September). $1000 mil. refinery plan for Wagerup!, *Western Herald*, p. 1.
9. Dames & Moore Consultancy. (1978). *Wagerup alumina project. Environmental review and management programme - Supplement*. Perth: Alcoa Australia.
10. Anon. (1976, 29th September). Council puts its view on refinery. *The West Australian*.
11. Anon. (1976, 10th December). $650m. alumina works plan for SW. *The West Australian*, p. 1&8.
12. Kellow, A., & Niemeyer, S. (1999). The development of environmental administration in Queensland and Western Australia: Why are they different? *Australian Journal of Political Science, 34*(2), 205–22.
13. Government of Western Australia & Alcoa of Australia Ltd. (1978). *Alumina refinery (Wagerup) agreement and acts amendment act 1978*. Perth: Government of Western Australia.
14. Environmental Protection Authority. (1978). *Wagerup alumina refinery proposal by Alcoa of Australia limited. Bulletin 50*. Perth: EPA.
15. Alcoa. (1978). *Wagerup alumina project environmental review and management programme*. Perth: Alcoa Australia.
16. WA Parliamentary Debates - Hansard. (2006). Standing committee on environment and public affairs - eleventh report - Alcoa refinery at Wagerup inquiry - Motion (pp. 2038a–42a). Perth: Parliament of Western Australia - Hansard.
17. Osborn, W. (Ed.). (2003). *Mr Wayne Osborn's opening statement to the committee, September 8 2003*. Perth.
18. Alcoa. (1997). *Annual Report*. Pittsburgh: Alcoa, Inc.
19. Ritchie, I. (1998, November). *The Changing Face of Extractive Metallurgy*. Paper presented at the Australian Academy of Technological Sciences and Engineering Symposium: Technology - Australia's future: New technology for traditional industries, Tokyo.
20. Standing Committee on Environment and Public Affairs. (2004). *Report of the Standing Committee on Environment and Public Affairs in relation to the Alcoa refinery at Wagerup inquiry*. Perth, Western Australia: Legislative Council.
21. Lippmann, M. (Ed.). (2009). *Environmental toxicants: human exposures and their health effect*. Hoboken, N.J.: Wiley & Sons.
22. Alcoa. (2007). Alumina Refining. Retrieved 1st May, 2007, from http://www.alcoa.com/australia/en/info_page/refining.asp

23. Southwell, M. (2001, 29th November). Cancer secret. *The West Australian*, pp. 1, 3.
24. Southwell, M. (2002, 27th June). Alcoa cancer rate alarm. *The West Australian*, p. 3.
25. Forster, P. (1990). *Internal memo*. Perth: Alcoa World Alumina.
26. Southwell, M. (2002, 24th May). Alcoa memo lists problem. *The West Australian*, p. 7.
27. Galton-Fenzi, B. (1997). *Liquor burning impact assessment (some health issues considered)*. Perth: The Healthy Worker Pty. Ltd.
28. Smith, S. (2002, 9th June). Alcoa's bad odour. *Radio National: Background Briefing*.
29. Drew, R. (1997). *An assessment of liquor burning odour emissions at Wagerup*. Melbourne: SHE Pacific Pty Ltd, Safety.
30. Mercer, A. (2001). *Report on Wagerup health survey*. Perth: Department of Public Health - University of Western Australia.
31. Monash University, & University of Western Australia. (2002). *Healthwise cancer and mortality study, first report*. Melbourne & Perth: MU & UWA.
32. Musk, A. W., & de Klerk, N. H. (2000). *Health effects from liquor burning unit emissions in an alumina refinery*. Perth: University of Western Australia.
33. Survey Research Centre. (2001). *Report on Wagerup health survey*. Perth: Department of Public Health - University of Western Australia.
34. Holman, D. (2002). *Summary and recommendations*. Perth: The Wagerup Medical Practitioners' Forum.
35. Utting, K. (2002, 19th March). People we do have a problem – Alcoa. *Harvey Reporter*.
36. Cullen, M. (2002). *Wagerup alumina refinery: health issues*. New Haven: Yale University.
37. Donoghue, A. M., & Cullen, M. R. (2007). Air emissions from Wagerup alumina refinery and community symptoms: An environmental case study. *Journal of Occupational and Environmental Medicine, 49*(9), 1027-1039.
38. Cook, M. (2003). *Six month report of Yarloop community health clinic*. Yarloop: South West Population Health Unit - Yarloop Community Clinic.
39. Healthwise. (2004). *Healthwise cancer incidence & mortality study*. Melbourne: Centre for Occupational and Environmental Health - Monash University.
40. Calhoun, R., Retallack, C., Christman, A., & Fernando, H. (2008). *Meteorological mechanisms and pathways of pollution exposure: Coherent doppler Lidar deployment in Wagerup. Final report*. Tempe, AZ: Arizona State University.
41. Department of Environment and Conservation. (2008). *Conditions of license (License Number 6217/12, File Number L80/83)*. Perth: DEC.
42. Flint, J. (2008, 27th May). Tests confirm town's fears. *The Sunday Times*, p. 24.
43. Holman, D. (2008). *The Wagerup and surrounds community health survey*. Perth: Telethon Institute for Child Research.

44. Alcoa. (2001). *Alcoa Wagerup land management draft proposal.* Perth: Alcoa World Alumina.
45. Alcoa. (2002). *Alcoa Wagerup land management revised proposal.* Perth: Alcoa World Alumina.
46. Alcoa. (2005). *Environmental review and management plan - Wagerup refinery unit three.* Perth: Alcoa World Alumina.
47. Alcoa. (2004). *Area B proposal. Letter to the Land Management Working Group.* Pinjarra: Alcoa World Alumina.
48. Walkington, P. (2005). *Environmental Protection Authority briefing note: Site visit on 7 July 20056 for the Alcoa Wagerup refinery unit 3 expansion.* Perth: EPA.
49. Community Alliance for Positive Solutions (CAPS) Inc. (2009). *Noise regulation 17 application - Alcoa Wagerup refinery. Letter to the Department of Environment and Conservation.* Yarloop: CAPS.
50. Ross, D. (2003). *Land management 2001-2003. Initial briefing paper.* Bunbury: Edith Cowan University.
51. Ross, D. (2003). *Reviewing the land management process: Some common ground at a point in the process. A report on the collaboration between Alcoa, Wagerup and Yarloop/Hamel property owners.* Bunbury: Edith Cowan University.
52. Flint, J. (2006, 10th September). Alcoa expands despite toxins. *The Sunday Times,* p. 16.
53. Alcoa. (2009). Community consultation. Retrieved 5th July, 2009, from http://www.alcoa.com/australia/en/info_page/wagerup_comm_consultation.asp
54. Welker Environmental Consultancy. (2003). *Western Australian licence conditions. Independent strategic review. Final report.* Perth: Department of Environmental Protection.
55. Department of Environment and Conservation. (2009). Wagerup Tripartite Group. Retrieved 4th April, 2009, from http://portal.environment.wa.gov.au/portal/page?_pageid=93,931490&_dad=portal&_schema=PORTAL
56. Alcoa. (2005, 8th February). Wagerup unit three consultation update. *Harvey Reporter,* p. 17.
57. Department of Environment and Conservation. (2009). Wagerup Tripartite Group meeting reports. Retrieved 4th April, 2009, from http://portal.environment.wa.gov.au/portal/page?_pageid=93,2384718&_dad=portal&_schema=PORTAL
58. Alcoa. (2005). *Environmental review and management programme Wagerup refinery unit three* Perth: Alcoa World Alumina.
59. Flint, J. (2006, 8th October). Surviving in the shadow of Alcoa. Town fears expansion. *The Sunday Times,* p. 50.
60. Community Alliance for Positive Solutions Inc. (2006). *Appeal to Environmental Protection Authority - Wagerup alumina refinery - Increase in production to 4.7 Mtpa; and Wagerup cogeneration plant assessment #1527.* Yarloop: CAPS.
61. Yarloop and Districts Concerned Residents' Group. (2005). *Submission against the Alcoa Wagerup refinery.* Yarloop: private copy.
62. Holman, D., Harper, A., Somers, M., Galton-Fenzi, B., & Phillips,

M. (2005). *Wagerup refinery Unit Three Expansion - Letter to the Environmental Protection Authority*. Perth: Independent Members of the Wagerup Medical Practitioners' Forum.
63. Environmental Protection Authority. (2006). *Wagerup alumina refinery - increase in production to 4.7 Mtpa; and Wagerup cogeneration plant. Bulletin 1215*. Perth: EPA.
64. McGowan, M. (2006). *Environmental approval for the Alcoa expansion. Statement by the Minister for the Environment*. Perth: Government of Western Australia.
65. Alcoa. (2008). Global financial crisis puts Wagerup 3 on hold. Retrieved 11th November, 2008, from http://www.alcoa.com/australia/en/news/releases/Wagerup3_on_hold.asp
66. WA Legislative Council. (2007). *Alumina refinery (Wagerup) agreement and acts amendment act 1978 - variation agreement - disallowance* (pp. 3176g–84a / 3171). Perth: Parliament of Western Australia - Hansard.
67. Llewellyn, P. (2008, 15th May). *Parliamentarians turn their backs on Yarloop residents - Media release*. Perth: The Greens.
68. Australian Bureau of Statistics. (1996). *Census 1996*. Canberra: ABS.
69. Australian Bureau of Statistics. (1996). *Community profiles for Harvey*. Canberra: ABS.
70. Bruce, B. (2008, 5th January). At death's door ... but fighting back. *The New Zealand Herald*. Retrieved 4th July 2008, from http://www.nzherald.co.nz/pollution/news/article.cfm?c_id=281&objectid=10485296&pnum=1.
71. Hepburn, H. (2007, 17th April). Yarloop's economy dwindles. *Hervey-Leschenault Reporter*, p. 1.
72. Australian Bureau of Statistics. (2002). *2001 census community profile series: Yarloop*. Canberra: ABS.
73. Australian Bureau of Statistics. (2006). *2006 census quickstats: Yarloop (state suburb)* Canberra: ABS.
74. Chartres, M., & Rowland, P. (2004). *Baseline valuation review*. Perth: Megaw & Hogg Incorporating Property Valuation & Consulting Services.
75. Walker, C. (2002). *On the valuation of properties. Wagerup alumina refinery & Yarloop townsite. III*. Roleystone: Geo & Hydro Environmental Management Pty Ltd.
76. Community Alliance for Positive Solutions (CAPS) Inc. (2009). About CAPS. Retrieved 3rd March, 2009, from http://www.caps6218.org.au/about.php
77. Flint, J. (2007, 5th August). Erin Brockovich takes on Alcoa. *The Sunday Times*.
78. ABC News. (2008, 3rd March). Erin Brockovich leads class action against Alcoa. *ABC*.
79. Business WA Today. (2009). Alcoa to fight pollution charge. Retrieved 23rd July, 2009, from http://business.watoday.com.au/business/alcoa-to-fight-pollution-charge-20090723-duhn.html
80. Australian Associated Press. (2009, 23rd July). Alcoa to face court over Wagerup charge. *Western Australian Business News [on-line]*. Retrieved 29.07.2009, from http://www.wabusinessnews.com.au/en-story.php?/1/73987/Alcoa-to-face-court-over-Wagerup-charge/

dba.
81. Stolley, G. (2009, 22nd December). Alcoa Wagerup trial date set. *The West Australian*. Retrieved 2nd January 2010, from http://au.news.yahoo.com/thewest/a/-/breaking/6615683/alcoa-wagerup-trial-date-set/.

CHAPTER 3

1. Holmes, D. (2008). *The Wagerup and surrounds community health survey*. Perth: Telethon Institute for Child Research.
2. Holmes, D., & Girardi, G. (1999). *Survey of productivity and environmental strategies in Australian manufacturing companies*. Melbourne: Australian Industry Group and Monash Centre for Environmental Management.
3. Standing Committee on Environment and Public Affairs. (2004). *Report of the Standing Committee on Environment and Public Affairs in relation to the Alcoa refinery at Wagerup inquiry*. Perth, Western Australia: Legislative Council.
4. Baum, F. (1999). The role of social capital in health promotion: Australian perspectives. *Health Promotion Journal of Australia, 9*(3), 171–9.
5. Cox, E. (Writer) (1995). A truly civil society [Radio]. In ABC (Producer), *The 1995 Boyer Lectures*: Radio National.
6. Portes, A. (1998). Social capital: Its origins and applications in modern sociology. *Annual review of sociology, 24*, 1–24.
7. Croft, J. (2005). *Results of interviews conducted in Yarloop*. Perth: Community Capacity Buidling Branch - Department of Local Government and Regional Development.
8. Alcoa. (2006). *Wagerup unit three - Overview of sustainability & permitting components*. Perth: Alcoa World Alumina.
9. Flint, J. (2006, 8th September). Tenants sign health forms. *The Sunday Times*, p. 12.
10. Ferguson, F. (2006, 22nd November). Yarloop exodus. *Harvey Mail*, p. 1.
11. Utting, K. (2006, 24th October). Hospital close after 111 years. *Harvey-Leschenault Reporter*, p. 1.
12. Buggins, A., Sadler, M., & Morfesse, L. (2006, 4th February). Health fears force police out. *The West Australian*, p. 4.
13. Utting, K. (2006, 14th February). Staff shift a blow for Yarloop. *Harvey-Leschenault Reporter*, p. 1.
14. Haines, A. (2006, 9th February). Police leave Yarloop. *Mandurah Mail*, p. 20.
15. Nobbs, J. (2003). *Yarloop public rental housing. Letter to K.J. Leece, Chief Executive Officer with the Department of Housing and Works*. Bunbury: Homeswest.
16. Australian Bureau of Statistics. (2007). *House Price Indexes: Eight Capital Cities, June 2007 quarter [Cat. No. 6416.0]*. Canberra: ABS.
17. Walker, C. (2002). *On the valuation of properties. Wagerup alumina refinery & Yarloop townsite. III*. Roleystone: Geo & Hydro Environmental Management Pty Ltd.

18. Arnstein, S. R. (1969). A ladder of citizen participation. *Journal of the American Institute of Planners, 35*(4), 216–24.
19. Alcoa. (2007). *WagerUpdate 23*. Perth: Alcoa World Alumina.

CHAPTER 4

1. Mayman, J. (2002, 11th-12th May). The stink of Uncle Al. *The Weekend Australian*, p. 19.
2. Singleton, J., & Elliot, P. (2003). *Comments on the review report (June 2003) of Alcoa's revised proposal (January, 2002). Correspondence provided to the Land Management Meetings by senior consultants of URS*. Perth: URS.
3. Standing Committee on Environment and Public Affairs. (2004). *Report of the Standing Committee on Environment and Public Affairs in relation to the Alcoa refinery at Wagerup inquiry*. Perth, Western Australia: Legislative Council.
4. Hahn, T. (2002). *In the loop. Video-recording*. Bunbury: Centre for Social Research - Edith Cowan University.
5. Ross, D. (2003). *Land management 2001-2003. Initial briefing paper*. Bunbury: Edith Cowan University.
6. Ross, D. (2003). *Reviewing the land management process: Some common ground at a point in the process. A report on the collaboration between Alcoa, Wagerup and Yarloop/Hamel property owners*. Bunbury: Edith Cowan University.
7. Ross, D. (2002). *Yarloop - Alcoa at the crossroads – naming the issues*. Bunbury: Centre for Social Research - Edith Cowan University.
8. Achbar, M., Abbott, J., & Bakan, J. (2004). *The corporation (Film)*. London: Metrodome Distributions.
9. Nelson, N., & Wright, S. (1995). *Power and Participatory Development: Theory and Practice*. London: Intermediate Technology.
10. Fook, J. (2002). *Social work: Critical theory and practice*. London: Sage Publications.
11. Foucault, M. (1980). *Power/knowledge: Selected interviews and other writings, 1972-1977 (C. Gordon, Trans.)*. Brighton: Harvester Press.
12. Smith, L. T. (1999). *Decolonizing methodologies: Research and indigenous peoples*. London: Zed Books.
13. Axelrod, R. (1997). *The complexity of co-operation*. Princeton: Princeton University Press.
14. Belenky, M. F., Clinchy, B., Goldberger, N., & Tarule, J. (1986). *Women's ways of knowing: The development of self, voice, and mind*. New York: Basic Books.
15. Goldberger, N. R., Tarule, J., Clinchy, B., & Belenky, M. F. (Eds.). (1996). *Knowledge, difference, and power: Essays inspired by women's ways of knowing*. New York: Basic Books.
16. Community Alliance for Positive Solutions (CAPS) Inc. (2009). The Yarloop bucket brigade. Retrieved 5th May, 2009, from http://www.caps6218.org.au/yarloop_bucket.php
17. Smith, D. E. (1990). *The conceptual practices of power: A feminist sociology of knowledge*. Boston: Northeastern University Press.
18. Leonard, P. (1997). *Postmodern welfare: Reconstructing an emancipatory project*. London Sage Publications.

19. Government of Western Australia. (2003). *Transcript of evidence, 8th September, 2003. Alcoa Wagerup Refinery at Wagerup.* Perth: Standing Committee on Environment and Public Affairs.
20. Alcoa. (2007). Wayne Osborne to retire from Alcoa of Australia. Retrieved 5th May, 2009, from http://www.alcoa.com/australia/en/news/releases/osborn_retire.asp
21. Alcoa. (2002). Visions, values: Principles. Retrieved 11th September, 2002, from http:/www.alcoa.com.au/governance/values.shtml
22. Alcoa. (2008). *Sustainability 08. Unlocking the solutions to sustainability.* Perth: Alcoa World Alumina.
23. Alcoa. (2002). *Alcoa Wagerup land management revised proposal.* Perth: Alcoa World Alumina.
24. Alcoa. (2002 / 2003). Wagerup community consultative network (Various citations in Alcoa's monthly report of network meetings). *The Harvey Reporter*
25. Utting, K. (2002, 19th March). People we do have a problem – Alcoa. *Harvey Reporter.*
26. Alcoa. (2003). Proud to be part of the community (Advertisement). *The Harvey Reporter.*
27. Bartlett, L., & Pownall, M. (2003, 27th February). Alcoa's operations in Australia: Interview with John Pizzey. *ABC Radio 720*
28. Sandman, P. (2002). Laundry list of 50 outrage reducers. Retrieved 7th November, 2003, from http://www.psandman.com/col/laundry.htm
29. Amalfi, C. (2003, 18th March). Judge accuses Alcoa. *The Western Australian,* p. 12.
30. Alcoa. (2004). Our future, your future. (Alcoa – Growing our future series of advertisements – several). *The Harvey Reporter.*
31. Alcoa. (2008). Alcoa invests in community future - Wagerup sustainability fund & Peel regional office. Retrieved 5th May, 2009, from http://www.alcoa.com/australia/en/news/releases/investing_in_community.asp
32. Alcoa. (2005). *WagerUpdate 12.* Perth: Alcoa World Alumina.
33. Alcoa. (2006). *Alcoa submission to the State Infrastructure Strategy.* Perth: Alcoa World Alumina.
34. Brueckner, M. (2008). *Balancing people, place and prosperity: Lessons from Western Australia.* Paper presented at the Symposium: Wealth and prosperity in the period of global transformation: experience of Russia and Asia.
35. Wilson, N. (2006, 22nd July). Can-do Alcoa pays a generous social dividend for new alumina plant. *The Australian.*
36. Towie, N. (2008, 14th December). Alcoa charged: Criminal negligence case to answer. *Sunday Times* p. 16.
37. Donoghue, A. M., & Cullen, M. R. (2007). Air emissions from Wagerup alumina refinery and community symptoms: An environmental case study. *Journal of Occupational and Environmental Medicine, 49*(9), 1027–39.

CHAPTER 5

1. Anderson, T. L., & Huggins, L. E. (2004). Economic growth--the essence of sustainable development. *Reason, 35*(8).
2. Beckerman, W. (2002). *A poverty of reason. Sustainable development and economic growth*. Oakland, CA: The Independent Institute.
3. United Nations - General Assembly. (2001). *Draft resolution introduced on ways to promote peace and progress*. New York: UN.
4. Mazur, J. (2000). Labor's new internationalism. *Foreign Affairs, 79*(1), 79-93.
5. Adams, R. H. J. (2002). *Economic growth, inequality and poverty: findings from a new data set*. Washington, DC: World Bank.
6. Grossman, G. M., & Krueger, A. B. (1995). Economic growth and the environment. *The Quarterly Journal of Economics, 110*(2), 353-77.
7. Walker, B., Carpenter, S., Anderies, J., Abel, N., Cumming, G., Janssen, M., et al. (2002). Resilience management in social-ecological systems: a working hypothesis for a participatory approach. *Conservation Ecology, 6*(1), 14.
8. Kellow, A., & Niemeyer, S. (1999). The development of environmental administration in Queensland and Western Australia: Why are they different? *Australian Journal of Political Science, 34*(2), 205-22.
9. Australian Bureau of Statistics. (2006). *Australian national accounts. State accounts [Cat. No. 5220.0]* (No. 5220.0). Canberra: ABS.
10. Kryger, T. (2007). *State economic and social indicators* (No. 14 2007-08). Canberra: Commonwealth of Australia.
11. Nicol, T. (2006). WA's mining boom: Where does it leave the environment? *ECOS, 133*(Oct-Nov), 12-3.
12. Aucoin, P. (1990). Administrative reform in public management: paradigms, principles, paradoxes and pendulums. *Governance, 3*(2), 115-37.
13. Battin, T. (1991). What is this thing called economic rationalism. *Australian Journal of Social Issues, 26*(5), 294-307.
14. Beeson, M., & Firth, A. (1998). Neoliberalism as a political rationality: Australian public policy since the 1980s. *Journal of Sociology, 34*(3), 215-31.
15. Bell, S. (1997). Globalisation, neoliberalism and the transformation of the Australian state. *Australian Journal of Political Science, 32*(3), 345-67.
16. Carroll, J., & Manne, R. (Eds.). (1992). *Shutdown: The Failure of Economic Rationalism and How to Rescue Australia*. Melbourne: Text Publishing Company.
17. Coleman, W., & Hagger, A. (2001). *Exasperating calculators*. Sydney: Macleay Press.
18. Henderson, D. (1995). The revival of economic liberalism: Australia in an international perspective. *The Australian Economic Review, 58* (1st Quarter), 59-85.
19. James, C., Jones, C., & Norton, A. (Eds.). (1993). *A Defence of Economic Rationalism*. Sydney: Allen & Unwin.
20. King, S., & Lloyd, P. (Eds.). (1993). *Economic rationalism: Dead end or way forward?* Sydney: Allen & Unwin.

21. Langmore, J., & Quiggin, J. (1994). *Work for all*. Melbourne: Melbourne University Press.
22. Orchard, L. (1998). Managerialism, Economic Rationalism and Public Sector Reform in Australia: Connections, Divergences, Alternatives. *Australian Journal of Public Administration, 57*(1), 19-32.
23. Painter, M. (1996). Economic policy, market liberalism and the 'end of Australian politics'. *Australian Journal of Political Science, 31*(3), 287-99.
24. Pusey, M. (1991). *Economic rationalism in Canberra: a nation-building state changes its mind*. Cambridge: Cambridge University Press.
25. Pusey, M. (2003). *The experience of middle Australia: the dark side of economic reform*. Cambridge: Cambridge University Press.
26. Rees, S., Rodley, G., & Stilwell, F. (Eds.). (1993). *Beyond the market. Alternatives to economic rationalism*. Leichhardt, NSW: Pluto Press.
27. Watson, I., Buchanan, J., Campbell, I., & Briggs. (2003). *Fragmented futures: new challenges in working life*. Annandale, NSW: Federation Press.
28. Brueckner, M. (2007). The Western Australian Regional Forest Agreement: economic rationalism and the normalisation of political closure. *Australian Journal of Public Administration, 66*(2), 148-58.
29. Churches, S. C. (2000). Courts and parliament dysfunctional in review: forest management as a case study of bureaucratic power. *Australian Journal of Administrative Law, 7*, 141-56.
30. Conacher, A., & Conacher, J. (2000). *Environmental Planning and Management in Australia*. Melbourne: Oxford University Press.
31. Barber, B. R. (1984). *Strong Democracy. Participatory Politics for a New Age*. Berkeley: University of California Press.
32. Walker, K. J. (2002). Uncertainty, epistemic communities and public policy. In J. W. Handmer, T. W. Norton & S. R. Dovers (Eds.), *Ecology, Uncertainty and Policy. Managing Ecosystems for Sustainability* (pp. 261-90). Harlow, UK: Pearson Education Ltd.
33. Brueckner, M. (2009). *Openness in the face of systemic constraints*. Cologne: Lampert Academic Publishing.
34. Aplin, G. (2000). Environmental rationalism and beyond: toward a more just sharing of power and influence. *Australian Geographer, 31*(3), 273-87.
35. Diesendorf, M., & Hamilton, C. (Eds.). (1997). *Human ecology, human economy. Ideas for an ecologically sustainable future*. St Leonards: Allen & Unwin.
36. Dovers, S. R. (2002). Policy, science, and community: interactions in a market-oriented world. *Australasian Journal of Ecotoxicology, 8*, 41-4.
37. Eckersley, R. (2001). Economic progress, social disquiet: the modern paradox. *Australian Journal of Public Administration, 60*(3), 89-97.
38. Gare, A. (2002). Human ecology and public policy: overcoming the hegemony of economics. *Democracy and Nature, 8*(1), 131-41.
39. Nevile, J. W. (1997). *Economic rationalism. Social philosophy masquerading as economic science. Background Paper No. 7*. Canberra: The Australia Institute.

40. State of the Environment Advisory Council. (1996). *Australia: state of the environment 1996*. Collingwood: CSIRO Publishing.
41. Mol, A. P. J., & Sonnenfeld, D. A. (2000). Ecological modernisation around the world: An introduction. *Environmental Politics, 9*(1), 1–14.
42. Murphy, J. (2000). Ecological modernisation. *Geoforum, 31*, 1–8.
43. Mol, A. P. J., & Spaargaren, G. (2000). Ecological modernisation theory in debate: A review. *Environmental Politics, 9*(1), 17–49.
44. Ackerman, F., & Gallagher, K. (2000). *Getting the prices wrong: The limits of market-based environmental policy. G-DAE Working Paper No. 00-05*. Medford, USA: Global Development and Environment Institute, Tufts University.
45. Costanza, R. (1987). Social traps and environmental policy. *BioScience, 37*(6), 407–12.
46. Hamilton, C. (2007). *Scorcher. The dirty politics of climate change*. Melbourne: Black Inc. Agenda.
47. Pearse, G. (2007). *High and dry: John Howard, climate change and the selling of Australia's future*. Camberwell, Vic.: Penguin.
48. WA Legislative Council. (2007). Alumina refinery (Wagerup) agreement and acts amendment act 1978 - variation agreement - disallowance (pp. 3176g–84a / 3171). Perth: Parliament of Western Australia - Hansard.
49. Turton, H. (2002). *The aluminium smelting industry: Structure, market power, subsidies and greenhouse gas emissions (Discussion Paper Number 44)*. Canberra: The Australia Institute.
50. Alcoa. (2005). *WagerUpdate 12*. Perth: Alcoa World Alumina.
51. Alcoa. (2007). *WagerUpdate 22*. Perth: Alcoa World Alumina.
52. Carpenter, A. (2006). *Brief ministerial statement: Alcoa Wagerup expansion*. Perth: Office of the Appeals Convenor.
53. Goot, M. (2002). Distrustful, disenchanted and disengaged? Public opinion on politics, politicians and the parties: An historical perspective. In D. Burchell & A. Leigh (Eds.), *The Prince's New Clothes: Why do Australians dislike their politicians* (pp. 9–46). Sydney: UNSW Press.
54. Jaensch, D. (1995). *Parliament, politics and people: Australian politics today* (2nd ed.). Melbourne: Longman Australia Pty Ltd.
55. Blind, P. K. (2007). *Building trust in government in the twenty-first century: Review of literature and emerging issues*. Paper presented at the 7th Global Forum on Reinventing Government.
56. Edelman. (2009). *Trust barometer*. Sydney: Edelman.
57. Whitmore, E., & Wilson, M. (2005). Popular resistance to global corporate rule: the role of social work (with a little help from Gramsci and Freire). In I. Ferguson, M. Lavalette & E. Whitmore (Eds.), *Globalisation, global justice and social work* (pp. 189–206). London: Routledge.
58. Department of State Development. (2009). *State agreements*. Perth: Government of Western Australia.
59. Standing Committee on Environment and Public Affairs. (2004). *Report of the Standing Committee on Environment and Public Affairs in relation to the Alcoa refinery at Wagerup inquiry*. Perth, Western Australia: Legislative Council.

60. Bureau of Transport and Regional Economics. (2006). *Skill shortages in Australia's regions.* Canberra: Commonwealth of Australia.
61. Government and Community Safety Industry Skills Council. (2009). *Government skills Australia - environmental scan.* Adelaide: Government Skills Australia.
62. Washburn, J. (2005). *University Inc. The corporate corruption of Amercian higher education.* Cambridge, MA: Basic Books.
63. Eisenhuth, S., & McDonald, W. (2007). *The writer's reader. Understanding journalism and non-fiction.* Cambridge: Cambridge University Press.

CHAPTER 6

1. McChesney, R. W. (1999). Introduction. In N. Chomsky (Ed.), *Profit over people: Neoliberalism and global order.* (pp. 7-18). New York: Seven Stories Press.
2. Pilger, J. (2006). *Freedom next time.* London: Bantam Press.
3. Bell, D. (1997). Anti-idyll: rural horror. In P. Cloke & J. Little (Eds.), *Contested countryside cultures: Otherness, marginalisation and rurality* (pp. 91-104). London: Routledge.
4. Pile, S., & Keith, M. (1997). *Geographies of resistance.* London Routledge.
5. Towie, N. (2009, 15th February). Alcoa let off DEC hook. *The Sunday Times,* p. 36.
6. Flyvbjerg, B. (1998). *Rationality and power: Democracy in practice. (Trans. S. Sampson).* Chicago: The University of Chicago Press.
7. Australian Association of Social Workers. (2002). *Code of ethics.* Barton: AASW.
8. Norgaard, R. (Ed.). (1994). *Development betrayed: The end of progress and a coevolutionary visioning of the future.* London: Routledge.
9. Alcoa. (2008). *Sustainability 08. Unlocking the solutions to sustainability.* Perth: Alcoa World Alumina.
10. MMSD Project. (2002). *Breaking new ground.* London: Earthscan.
11. Ross, D. (2009). Emphasizing the 'social' in corporate social responsibility: A social work perspective. In S. Idowu & W. Leal Filho (Eds.), *Professional perspectives of corporate social responsibility* (pp. 301-18). Heidelberg: Springer.
12. Mander cited in Hinkley, R. (2000). Developing corporate conscience. In S. Rees & S. Wright (Eds.), *Human rights, corporate responsibility: A dialogue* (pp. 287-95). Annandale: Pluto Press Australia.
13. Smallegange cited in Hahn, T. (2002). *In the loop. Video-recording.* Bunbury: Centre for Social Research - Edith Cowan University.
14. Standing Committee on Environment and Public Affairs. (2004). *Report of the Standing Committee on Environment and Public Affairs in relation to the Alcoa refinery at Wagerup inquiry.* Perth, Western Australia: Legislative Council.
15. Holthouse cited in Ross, D. (2003). *The business case for an ethic of love: The Alcoa/ECU partnership – informing theoretical and ethical constructs.* Bunbury: Centre for Regional Development & Research - Edith Cowan University.

16. Hartman, A. (1990). Many ways of knowing. *Social Work, January*, 3-4.
17. Mills, C. W. (1959). *The sociological imagination*. New York: Oxford University Press.
18. Hemmati, M. (2002). *Multi-stakeholders processes for governance and sustainability – beyond deadlock and conflict*. London: Earthscan.
19. Government of Western Australia. (2003). *Hope for the future: The Western Australian state sustainability strategy*. Perth: Department of the Premier and Cabinet.
20. Alcoa. (2006). *Summary response to appeals issues. Wagerup unit 3 expansion project*. Applecross: Alcoa World Alumina.
21. Keith, M., & Pile, S. (1993). *Place and the politics of identity*. London Routledge.
22. Flyvjberg, B. (2001). *Making social science matter: Why social inquiry fails and how it can succeed again*. Cambridge: Cambridge University Press.
23. Achbar, M., Abbott, J., & Bakan, J. (2004). *The corporation (Film)*. London: Metrodome Distributions.
24. Ife, J., & Tesorieroe, F. (2006). *Community development* (3rd ed.). French's Forest: Longman.
25. Pattenden, C. (2004). Sustaining community in a closed mining town. In D. Woods & G. Pass (Eds.), *Alchemies: Community exChanges*. Perth: Black Swan Press.
26. Young, I. M. (1990). *Justice and the politics of difference*. Princeton: Princeton University Press.
27. Ross, D. (2003). *Reviewing the land management process: Some common ground at a point in the process. A report on the collaboration between Alcoa, Wagerup and Yarloop/Hamel property owners*. Bunbury: Edith Cowan University.
28. Freire, P. (1970). *Pedagogy of the oppressed*. New York: Herder and Herder.
29. Ross, D. (2002). *Yarloop - Alcoa at the crossroads – naming the issues*. Bunbury: Centre for Social Research - Edith Cowan University.
30. Ross, D. (2007). A dis/integrative, dialogic model for social work education. *Social Work Education, 26*(5), 1-5.
31. Tennyson, R., & Wilde, L. (2000). *The guiding hand: Brokering partnerships for sustainable development*. Geneva: United Nations Office of Public Information.
32. Singleton, J., & Elliot, P. (2003). *Comments on the review report (June 2003) of Alcoa's revised proposal (January, 2002). Correspondence provided to the Land Management Meetings by senior consultants of URS*. Perth: URS.
33. Alcoa. (2002 / 2003). Wagerup community consultative network (Various citations in Alcoa's monthly report of network meetings). *The Harvey Reporter*
34. Chetkow-Yanoov, B. (1997). *Social work approaches to conflict resolution*. New York: Hawthorn Press.
35. Lagan, A. (2000). *Why ethics matter: Business ethics for business people*. Melbourne: Information Australia.
36. Curtin University of Technology. (2010). Alcoa Research Centre for

Stronger Communities. Retrieved 15th January, 2010, from http://strongercommunities.curtin.edu.au/research_new.htm.
37. Zadec, S. (2001). *The civil corporation: The new economy of corporate citizenship*. London Earthscan
38. Fox, C., & Miller, H. (1995). *Postmodern public administration: Toward discourse*. Sage Publications.
39. Dryzek, J. (1990). *Discursive democracy: Politics, policy and political science*. Cambridge: Cambridge University Press.
40. Saul, J. R. (2005). *The collapse of globalism and the reinvention of the world*. New York: The Overlook Press.
41. Thompson, N. (1998). *Promoting equality: Challenging discrimination and oppression in the human services*. London: Palgrave Macmillan
42. Foucault, M. (1980). *Power/knowledge: Selected interviews and other writings, 1972-1977 (C. Gordon, Trans.)*. Brighton: Harvester Press.
43. Alcoa. (2003, 24th September). Community consultative network. *Harvey Reporter*, p. 2.
44. McDermott, Q. (2005, 3rd October). Something in the air - ABC Four Corners.

CHAPTER 7

1. Flyvjberg, B. (2001). *Making social science matter: Why social inquiry fails and how it can succeed again*. Cambridge: Cambridge University Press.
2. Sandman, P. (2002). Laundry list of 50 outrage reducers. Retrieved 7th November, 2003, from http://www.psandman.com/col/laundry.htm
3. Donoghue, A. M., & Cullen, M. R. (2007). Air emissions from Wagerup alumina refinery and community symptoms: An environmental case study. *Journal of Occupational and Environmental Medicine, 49*(9), 1027–39.
4. Flint, J. (2006, 10th September). Alcoa expands despite toxins. *The Sunday Times*, p. 16.
5. Carson, R. (1962). *Silent spring*. New York: Penguin Books.
6. Freire, P. (1970). *Pedagogy of the oppressed*. New York: Herder and Herder.
7. Pettman, J. (1992). *Living in the margins: Racism, sexism and feminism in Australia*. North Sydney: Allen & Unwin.
8. Ross, D. (2009). Emphasizing the 'social' in corporate social responsibility: A social work perspective. In S. Idowu & W. Leal Filho (Eds.), *Professional perspectives of corporate social responsibility* (pp. 301–18). Heidelberg: Springer.
9. Hamann, R. (2009). An academic's perspective on the role of academics in corporate responsibility. In S. Idowu & W. Leal Filho (Eds.), *Professional perspectives of corporate social responsibility* (pp. 347–62). Heidelberg: Springer.
10. Government of Western Australia. (2003). *Hope for the future: The Western Australian state sustainability strategy*. Perth: Department of the Premier and Cabinet.
11. Alcoa. (2002). Visions, values: Principles. Retrieved 11th September,

2002, from http:/www.alcoa.com.au/governance/values.shtml
12. Brueckner, M. (2008). *Balancing people, place and prosperity: Lessons from Western Australia*. Paper presented at the Symposium: Wealth and prosperity in the period of global transformation: experience of Russia and Asia.
13. Hunold, C., & Young, I. M. (1998). Justice, democracy and hazardous siting. *Political Studies, 46*, 82–95.
14. Young, I. M. (1990). *Justice and the politics of difference*. Princeton: Princeton University Press.
15. Earles, W. (1999). *Powerless places and placeless powers: The (re) shaping of human services institutions*. University of Western Australia., Perth.
16. *Cameron B. Auxer et al. vs. ALCOA, INC*, (2009).
17. Amaeshi, K. M., & Adi, B. (2007). Reconstructing the corporate social responsibility construct in Utlish. *Business Ethics: A European Review, 16*(1), 3–18.
18. Korhonen, J. (2002). The dominant economics paradigm and corporate social responsibility. *Corporate Social Responsibility and Environmental Management, 9*(1), 66–79.
19. Banerjee, B. S. (2007). *Corporate social responsibility: The good, the bad and the ugly*. Cheltenham, UK: Edward Elgar.
20. Brueckner, M., & Mamun, M. A. (in print). Living downwind from CSR: A community perspective on corporate practice. *Business Ethics: A European Review*.
21. Alcoa. (2009). *Sustainability 08. Unlocking the solutions to sustainability*. Perth: Alcoa World Alumina.
22. Reich, R. (2008). *Supercapitalism. The transformation of business, democracy and everyday life*. Melbourne: Scribe.
23. Achbar, M., Abbott, J., & Bakan, J. (2004). *The corporation (Film)*. London: Metrodome Distributions.
24. Carroll, A. B. (1991). The pyramid of corporate social responsibility: towards the moral management of organizational stakeholders *Business Horizons July-August*, 39–48.
25. Lantos, G. P. (2001). The boundaries of strategic corporate social responsibility. *Journal of Consumer Marketing, 18*(7), 595–630.
26. Orlitzky, M., Schmidt, F. L., & Rynes, S. L. (2003). Corporate social and financial performance: A meta-analysis. *Organization Studies, 24*(3), 403–41.
27. Agle, B. R., Mitchell, R. K., & Sonnenfeld, J. A. (1999). Who matters to CEOs? An Investigation of stakeholder attributes and salience, corporate performance, and CEO values. *Academy of Management Journal, 42*(5), 507–25.
28. Mitchell, R. K., Agle, B. R., & Wood, D. J. (1997). Toward a Theory of Stakeholder Identification and Salience: Defining the Principle of Who and What Really Counts. *The Academy of Management Review, 22*(4), 853–86.
29. Benn, S., & Dunphy, D. (2007). *Corporate governance and sustainability. Challenges for theory and practice*. New York: Routledge.
30. Banerjee, S. B. (2001). Managerial perceptions of corporate environmentalism. *Journal of Management Studies, 38*(4), 489–513.

31. Elkington, J., & Fennell, S. (1998). Partners for sustainability. *Greener Management International, 24*, 48–60.
32. Valentine, T. (1996). Economic rationalism vs the entitlement consensus. *Policy*(Spring), 3–10.
33. Saul, J. R. (1997). *The unconscious civilisation*. Ringwood (VIC): Penguin Books Australia Ltd.
34. Hobbs, A. (2008, 15th September). Lobby groups urge Barnett to get moving. *The West Australian*.
35. Carpenter, A. (2006). *Brief ministerial statement: Alcoa Wagerup expansion*. Perth: Office of the Appeals Convenor.
36. Kellow, A., & Niemeyer, S. (1999). The development of environmental administration in Queensland and Western Australia: Why are they different? *Australian Journal of Political Science, 34*(2), 205–22.
37. Walker, B., Carpenter, S., Anderies, J., Abel, N., Cumming, G., Janssen, M., et al. (2002). Resilience management in social-ecological systems: a working hypothesis for a participatory approach. *Conservation Ecology, 6*(1), 14.
38. Brueckner, M. (2007). The Western Australian Regional Forest Agreement: economic rationalism and the normalisation of political closure. *Australian Journal of Public Administration, 66*(2), 148–58.
39. Phillips, H. C. J., & Kerr, L. (2005). Western Australia. *Australian Journal of Politics & History, 51*(2), 297–302.
40. Environmental Protection Authority. (2007). *State of the Environment 2007 Western Australia*. Perth: EPA.
41. Hinwood, A. (2007). *EPA media statement - EPA reveals Western Australia's environment report card*. Perth: EPA.
42. Lawson, R. (2009, 14th September). Gorgon approved, valued at $43bn. *WA Business News*, from www.wsbusinessnews.com.au/story/1/75230/gorgon-approved-valued-at-43bn.
43. Daly, H. E., & Cobb, J. B. (1989). *For the Common Good: Redirecting the Economy Towards the Community, the Environment and a Sustainable Future*. Boston: Beacon Press.
44. Handmer, J. W., Dovers, S. R., & Norton, T. W. (2001). Managing ecosystems for sustainability: challenges and opportunities. In J. W. Handmer, T. W. Norton & S. R. Dovers (Eds.), *Ecology, Uncertainty and Policy. Managing Ecosystems for Sustainability* (pp. 291–303). Harlow, UK: Pearson Education Ltd.
45. Nevile, J. W. (1997). *Economic rationalism. Social philosophy masquerading as economic science. Background Paper No. 7*. Canberra: The Australia Institute.
46. van Bavel, R., & Gaskell, G. (2004). Narrative and systemic modes of thinking. *Culture and Psychology, 10*(4), 418–39.
47. Hamilton, C. (1996). Generational Justice: The marriage of sustainability and social equity. *Australian Journal of Environmental Management, 3*(3), 163–73.
48. Meppem, T., & Gill, R. (1998). Planning for sustainability as a learning concept. *Ecological Economics, 26*, 121–37.
49. Hamilton, C. (2002). Dualism and sustainability. *Ecological Economics, 42*(1-2), 89–99.
50. Pemberton, J. (1988). 'The end' of economic rationalism. *Australian*

Quarterly, 60, 188-99.
51. Hamilton, C., & Maddison, S. (2007). *Silencing dissent. How the Australian government is controlling public opinion and stifling debate*. Crows Nest, NSW: Allen & Unwin.
52. Stacey, R. D. (1993). *Strategic management and organisational dynamics*. London: Pitman.
53. Mitroff, I. (1998). *Smart thinking for crazy times: the art of solving the right problem*. San Francisco: Berrett-Koehler.
54. Turton, H. (2002). *The aluminium smelting industry: Structure, market power, subsidies and greenhouse gas emissions (Discussion Paper Number 44)*. Canberra: The Australia Institute.
55. ABC News. (2008, 5th July). Lead contamination prompts boost to DEC. *ABC*, from Online: http://www.abc.net.au/news/stories/2007/11/29/2105423.htm.
56. Standing Committee on Education and Health. (2007). *Inquiry into the cause and extent of lead pollution in the Esperance area* Perth: Western Australia. Parliament. Legislative Assembly.
57. Taylor, R. (2007, 6th September). Esperance lead report scathing of DEC. *The West Australian*.
58. Saul, J. R. (2005). *The collapse of globalism and the reinvention of the world*. New York: The Overlook Press.
59. Foucault, M. (1997). Technologies of the self. In P. Rabinow (Ed.), *Michel Foucault. Ethics: subjectivity and truth* (pp. 223-52). New York: The New Press.
60. Smith, D. E. (1990). *The conceptual practices of power: A feminist sociology of knowledge*. Boston: Northeastern University Press.
61. hooks, b. (1994). *Outlaw culture: Resisting representations*. New York: Routledge.
62. hooks, b. (2000). *All about love*. London: New Visions.
63. Stanley, L., & Wise, S. (1993). *Breaking out again: Feminist ontology and epistemology* (2nd ed.). London: Routledge.
64. Leonard, P. (1997). *Postmodern welfare: Reconstructing an emancipatory project*. London Sage Publications.
65. Ife, J., & Tesorieroe, F. (2006). *Community development* (3rd ed.). French's Forest: Longman.
66. Raffensperger, C., & Tickner, J. A. (Eds.). (1999). *Protecting public health & the environment*. Washington, DC: Island Press.
67. Tickner, J. A. (2003). Precaution, environmental science, and preventive public policy. *New Solutions: A Journal of Environmental and Occupational Health Policy, 13*(3), 275-82.
68. Bailey, P., Yearley, S., & Forrester, J. (1999). Involving the public in local air pollution assessment: a citizen participation case study. *International Journal of Environment and Pollution, 11*(3), 290-303.
69. Funtowicz, S., & Ravetz, R. (1991). A new scientific methodology for global environmental issues. In R. Costanza (Ed.), *Ecological Economics*. New York: Columbia University Press.
70. Holman, H. R., & Dutton, D. B. (1978). A case for public participation in science policy formation and practice. *Southern California Law Review, 51*, 1505-34.
71. Krimsky, S. (1984). Beyond technocracy: new routes for citizen

involvement in social risk assessment. In J. C. Petersen (Ed.), *Citizen Participation in Science Policy* (pp. 43-61). Amherst: University of Massachusetts Press.
72. Laird, F. (1993). Participatory analysis, democracy, and technological decision making. *Science Technology & Human Values, 18*(3), 341-61.
73. Renn, O. (1992). Risk communication: towards a rational discourse with the public. *Journal of Hazardous Materials, 29*, 465-519.
74. Brueckner, M., & Horwitz, P. (2005). The use of science in environmental policy: A case study of the Regional Forest Agreement process in Western Australia. *Sustainability: Science, Practice, & Policy [On-line Journal]*. URL: http://ejournal.nbii.org/archives/vol1iss2/0412-017.brueckner.pdf, *1*(2), Accessed 5th July 2005.
75. Friedman, M. (1970, 13th September). The social responsibility of business is to increase its profits. *New York Times Magazine*, pp. 122-6.
76. Jensen, M. C. (2002). Value maximization, stakeholder theory, and the corporate objective function. *Business Ethics Quarterly, 12*(2), 235-56.
77. Wallich, H. C., & McGowan, J. J. (1970). Stockholder interest and the corporation's role in social policy. In W. J. Baumol (Ed.), *A new rationale for corporate social policy* (pp. 39-59). New York: Committee for Economic Development.
78. Kok, P., van der Wiele, T., McKenna, R., & Brown, A. (2001). A corporate social responsibility audit within a quality management framework. *Journal of Business Ethics, 31*(4), 285-97.
79. Banerjee, S. B. (2006). The problem with corporate social responsibility. In S. Clegg & C. Rhodes (Eds.), *Management ethics: What is to be done?* (pp. 55-76). New York: Palgrave.
80. Blowfield, M. (2005). Corporate social responsibility - the failing discipline and why it matters for international relations. *International Relations, 19*(2), 173-91.
81. Beder, S. (1997). *Global Spin. The Corporate Assault On Environmentalism*. Foxhole, UK: Green Books Ltd.
82. Beder, S. (2000). Hijacking sustainable development: a critique of corporate environmentalism. *Chain Reaction, 8*, 8-10.
83. Korten, D. (1995). *When corporations rule the world*. San Francisco: Barrett-Koehler Publishers.
84. Frankental, P. (2001). Corporate social responsibility - a PR invention? *Corporate Communications: An International Journal, 6*(1), 12-23.
85. Marsden, C. (2000). The new corporate citizenship of big business: part of the solution to sustainability. *Business and Society Review, 105*(1), 9-25.
86. McKenna, R., & Tsahuridu, E. (2001). Must managers leave ethics at home? Economics and moral anomie in business organisations. *Reason in Practice, 1*(3), 67-76.
87. Cox, K. (Ed.). (1997). *Spaces of globalization. Reasserting the power of the local*. New York: The Guilford Press.
88. Dovers, S. (2002). Sustainability: reviewing Australia's progress,

1992-2002. *International Journal of Environmental Studies, 59*(5), 559-71.
89. Doyle, T., & Kellow, A. (1995). *Environmental politics and policy making in Australia*. Melbourne: Macmillan.
90. Mercer, D., & Marden, P. (2006). Ecologically sustainable development in a 'quarry' economy: One step forward, two steps back. *Geographical Research, 44*(2), 183–203.
91. Australian State of the Environment Committee. (2001). *Australia State of the Environment 2001: Independent Report to the Commonwealth Minister for the Environment and Heritage*. Canberra: CSIRO Publishing on behalf of the Department of the Environment and Heritage.
92. Randal, A. (2008). Is Australia on a sustainability path? Interpreting the clues. *The Australian Journal of Agricultural and Resource Economics, 52*, 77–95.
93. Walker, K. J. (2001). Uncertainty, epistemic communities and public policy. In J. W. Handmer, T. W. Norton & S. R. Dovers (Eds.), *Ecology, Uncertainty and Policy. Managing Ecosystems for Sustainability* (pp. 262–90). Harlow, UK: Pearson Education Ltd.
94. O'Riordan, T., & Stoll-Kleemann, S. (2002). Deliberative democracy and participatory biodiversity. In T. O'Riordan & S. Stoll-Kleemann (Eds.), *Biodiversity, Sustainability and Human Communities. Protecting Beyond the Protected* (pp. 87-112). Cambridge: Cambridge University Press.
95. Brueckner, M., Duff, J., McKenna, R., & Horwitz, P. (2006). What are they taking us for? The participatory nature of Western Australia's Regional Forest Agreement process. *Australasian Journal of Environmental Management, 13*(1), 6–16.
96. Giddens, A. (1994). *Beyond Left and Right*. Cambridge: Polity Press.
97. Sztompka, P. (1999). *Trust: A Sociological Theory*. Cambridge: Cambridge University Press.
98. Brueckner, M. (2009). *Openness in the face of systemic constraints*. Cologne: Lampert Academic Publishing.
99. Dryzek, J. (1990). *Discursive democracy: Politics, policy and political science*. Cambridge: Cambridge University Press.
100. Community Alliance for Positive Solutions (CAPS) Inc. (2009). About CAPS. Retrieved 3rd March, 2009, from http://www.caps6218.org.au/about.php
101. Mills, C. W. (1959). *The sociological imagination*. New York: Oxford University Press.
102. Alcoa. (2009). *Wagerup refinery oxalate kiln recommissioning project*. Perth: Alcoa World Alumina.
103. Department of Environment and Conservation. (2007). Alcoa Wagerup licence renewal application - 6217/10 - Letter to Community Alliance for Positive Solutions.
104. Minister for Planning; Culture and the Arts. (2009, 4th June). Positive solutions for Yarloop and districts - Letter to the Community Alliance for Positive Solutions.
105. Department of Environment and Conservation. (2006). *Memorandum of understanding between Department of Environment*

and Conservation and Community Alliance for Positive Solutions for comparison of ambient air sampling and analysis techniques Perth: DEC.
106. Community Alliance for Positive Solutions (CAPS) Inc. (2009). Letter to Minister for Planning, Culture and the Arts.
107. Department of Environment and Conservation. (2009, 23rd January). Wagerup tripartite group - Letter to CAPS.
108. Minister for Environment; Youth. (2009, 26th August). Letter to Community Alliance for Positive Solutions.
109. Wynne, T. (2009). *Letter to the Community Alliance for Positive Solutions Inc*. Perth: Department of Environment and Conservation.
110. Community Alliance for Positive Solutions (CAPS) Inc. (2008, 18th December). List of conditions - Letter to Department of Environment and Conservation.
111. Community Alliance for Positive Solutions (CAPS) Inc. (2010). *Personal communication*. Yarloop.

REFERENCES

ABC News. (2008a, 3rd March). Erin Brockovich leads class action against Alcoa. *ABC*.

ABC News. (2008b, 5th July). Lead contamination prompts boost to DEC. *ABC*, from Online: http://www.abc.net.au/news/stories/2007/11/29/2105423.htm.

Abramowitz, M. (1989). *Thinking about growth*. Cambridge: Cambridge University Press.

Achbar, M., Abbott, J. & Bakan, J. (2004). *The corporation (Film)*. London: Metrodome Distributions.

Ackerman, F. & Gallagher, K. (2000). *Getting the prices wrong: The limits of market-based environmental policy. G-DAE Working Paper No. 00-05*. Medford, USA: Global Development and Environment Institute, Tufts University.

Adams, R. H. J. (2002). *Economic growth, inequality and poverty: findings from a new data set*. Washington, DC: World Bank.

Adams, W. M. (2006). *The future of sustainability. Re-thinking environment and development in the twenty-first century*. Gland, Switzerland: IUCN.

Agle, B. R., Mitchell, R. K. & Sonnenfeld, J. A. (1999). Who matters to CEOs? An Investigation of stakeholder attributes and salience, corporate performance, and CEO values. *Academy of Management Journal, 42*(5), 507–25.

Alcoa. (1978). *Wagerup alumina project environmental review and management programme*. Perth: Alcoa Australia.

Alcoa. (1997). *Annual Report*. Pittsburgh: Alcoa, Inc.

Alcoa. (2001). *Alcoa Wagerup land management draft proposal*. Perth: Alcoa World Alumina.

Alcoa. (2002a). *Alcoa Wagerup land management revised proposal*. Perth: Alcoa World Alumina.

Alcoa. (2002b). Visions, values: Principles. Retrieved 11th September, 2002, from http:/www.alcoa.com.au/governance/values.shtml

Alcoa. (2002 / 2003). Wagerup community consultative network (Various citations in Alcoa's monthly report of network meetings). *The Harvey Reporter*

Alcoa. (2003a, 24th September). Community consultative network. *Harvey Reporter*, p. 2.

Alcoa. (2003b). Proud to be part of the community (Advertisement). *The Harvey Reporter*

Alcoa. (2004a). *Area B proposal. Letter to the Land Management Working Group*. Pinjarra: Alcoa World Alumina.

Alcoa. (2004b). Our future, your future. (Alcoa – Growing our future series of advertisements – several). *The Harvey Reporter*.

Alcoa. (2005a). *Environmental review and management plan - Wagerup refinery unit three*. Perth: Alcoa World Alumina.

Alcoa. (2005b). *Environmental review and management programme Wagerup refinery unit three* Perth: Alcoa World Alumina.

Alcoa. (2005c, 8th February). Wagerup unit three consultation update. *Harvey Reporter,* p. 17.
Alcoa. (2005d). *WagerUpdate 12.* Perth: Alcoa World Alumina.
Alcoa. (2005e). *Your future our future.* Perth: Alcoa World Alumina.
Alcoa. (2006a). *Alcoa submission to the State Infrastructure Strategy.* Perth: Alcoa World Alumina.
Alcoa. (2006b). Green light for alumina refinery expansion. Retrieved 1st May, 2007, from http://www.alcoa.com/australia/en/info_page/WAG_home.asp
Alcoa. (2006c). *Summary response to appeals issues. Wagerup unit 3 expansion project.* Applecross: Alcoa World Alumina.
Alcoa. (2006d). *Wagerup unit three - Overview of sustainability & permitting components.* Perth: Alcoa World Alumina.
Alcoa. (2007a). Alcoa has been sustainably mining, refining and smelting in Australia since 1961. Retrieved 1st May, 2007, from http://www.alcoa.com/australia/en/alcoa_australia/australia_overview.asp
Alcoa. (2007b). Alumina Refining. Retrieved 1st May, 2007, from http://www.alcoa.com/australia/en/info_page/refining.asp
Alcoa. (2007c). Sustainability Approach. Retrieved 1st May, 2007, from http://www.alcoa.com/global/en/about_alcoa/sustainability/home_sustainability_approach.asp
Alcoa. (2007d). *WagerUpdate 22.* Perth: Alcoa World Alumina.
Alcoa. (2007e). *WagerUpdate 23.* Perth: Alcoa World Alumina.
Alcoa. (2007f). Wayne Osborne to retire from Alcoa of Australia. Retrieved 5th May, 2009, from http://www.alcoa.com/australia/en/news/releases/osborn_retire.asp
Alcoa. (2008a). Alcoa invests in community future - Wagerup sustainability fund & Peel regional office. Retrieved 5th May, 2009, from http://www.alcoa.com/australia/en/news/releases/investing_in_community.asp
Alcoa. (2008b). Global financial crisis puts Wagerup 3 on hold. Retrieved 11th November, 2008, from http://www.alcoa.com/australia/en/news/releases/Wagerup3_on_hold.asp
Alcoa. (2008c). *Sustainability 08. Unlocking the solutions to sustainability.* Perth: Alcoa World Alumina.
Alcoa. (2009a). Community consultation. Retrieved 5th July, 2009, from http://www.alcoa.com/australia/en/info_page/wagerup_comm_consultation.asp
Alcoa. (2009b). *Sustainability 08. Unlocking the solutions to sustainability.* Perth: Alcoa World Alumina.
Alcoa. (2009c). *Wagerup refinery oxalate kiln recommissioning project.* Perth: Alcoa World Alumina.
Amaeshi, K. M. & Adi, B. (2007). Reconstructing the corporate social responsibility construct in Utlish. *Business Ethics: A European Review, 16*(1), 3-18.
Amalfi, C. (2003, 18th March). Judge accuses Alcoa. *The Western Australian,* p. 12.
Anderson, T. L. & Huggins, L. E. (2004). Economic growth--the essence of sustainable development. *Reason, 35*(8).
Anon. (1976a, 10th December). $650m. alumina works plan for SW. *The West Australian,* p. 1&8.

Anon. (1976b, 29th September). Council puts its view on refinery. *The West Australian*.

Aplin, G. (2000). Environmental rationalism and beyond: toward a more just sharing of power and influence. *Australian Geographer, 31*(3), 273–87.

Arnstein, S. R. (1969). A ladder of citizen participation. *Journal of the American Institute of Planners, 35*(4), 216–24.

Aucoin, P. (1990). Administrative reform in public management: paradigms, principles, paradoxes and pendulums. *Governance, 3*(2), 115–37.

Australian Aluminium Council. (2004). *Sustainability report 2004*. Canberra: AAC.

Australian Associated Press. (2009, 23rd July). Alcoa to face court over Wagerup charge. *Western Australian Business News [on-line]*. Retrieved 29.07.2009, from http://www.wabusinessnews.com.au/en-story.php?/1/73987/Alcoa-to-face-court-over-Wagerup-charge/dba.

Australian Association of Social Workers. (2002). *Code of ethics*. Barton: AASW.

Australian Bureau of Statistics. (1996a). *Census 1996*. Canberra: ABS.

Australian Bureau of Statistics. (1996b). *Community profiles for Harvey*. Canberra: ABS.

Australian Bureau of Statistics. (2002). *2001 census community profile series: Yarloop*. Canberra: ABS.

Australian Bureau of Statistics. (2006a). *2006 census quickstats: Yarloop (state suburb)* Canberra: ABS.

Australian Bureau of Statistics. (2006b). *Australian national accounts. State accounts [Cat. No. 5220.0]* (No. 5220.0). Canberra: ABS.

Australian Bureau of Statistics. (2006c). *Measures of Australia's progress 2006* (No. 1370.0). Canberra: ABS.

Australian Bureau of Statistics. (2007). *House Price Indexes: Eight Capital Cities, June 2007 quarter [Cat. No. 6416.0]*. Canberra: ABS.

Australian State of the Environment Committee. (2001). *Australia State of the Environment 2001: Independent Report to the Commonwealth Minister for the Environment and Heritage*. Canberra: CSIRO Publishing on behalf of the Department of the Environment and Heritage.

Axelrod, R. (1997). *The complexity of co-operation*. Princeton: Princeton University Press.

Bailey, P., Yearley, S. & Forrester, J. (1999). Involving the public in local air pollution assessment: a citizen participation case study. *International Journal of Environment and Pollution, 11*(3), 290–303.

Banerjee, B. S. (2007). *Corporate social responsibility: The good, the bad and the ugly*. Cheltenham, UK: Edward Elgar.

Banerjee, S. B. (2001). Managerial perceptions of corporate environmentalism. *Journal of Management Studies, 38*(4), 489–513.

Banerjee, S. B. (2006). The problem with corporate social responsibility. In S. Clegg & C. Rhodes (Eds.), *Management ethics: What is to be done?* (pp. 55–76). New York: Palgrave.

Barber, B. R. (1984). *Strong Democracy. Participatory Politics for a New Age*. Berkeley: University of California Press.

Bartlett, L. & Pownall, M. (2003, 27th February). Alcoa's operations in Australia: Interview with John Pizzey. *ABC Radio 720*

Basagio, A. D. (1995). Methods of defining sustainability. *Sustainable Development, 3*, 109-119.

Battin, T. (1991). What is this thing called economic rationalism. *Australian Journal of Social Issues, 26*(5), 294-307.

Baum, F. (1999). The role of social capital in health promotion: Australian perspectives. *Health Promotion Journal of Australia, 9*(3), 171–9.

Beard, J. S., Chapman, A. R. & Gioia, P. (2000). Species richness and endemism in the Western Australian flora. *Journal of Biogeography, 27*, 1257–68.

Beckerman, W. (2002). *A poverty of reason. Sustainable development and economic growth*. Oakland, CA: The Independent Institute.

Beder, S. (1997). *Global Spin. The Corporate Assault On Environmentalism*. Foxhole, UK: Green Books Ltd.

Beder, S. (2000). Hijacking sustainable development: a critique of corporate environmentalism. *Chain Reaction, 8*, 8–10.

Beeson, M. & Firth, A. (1998). Neoliberalism as a political rationality: Australian public policy since the 1980s. *Journal of Sociology, 34*(3), 215–31.

Belenky, M. F., Clinchy, B., Goldberger, N. & Tarule, J. (1986). *Women's ways of knowing: The development of self, voice, and mind*. New York: Basic Books.

Bell, D. (1997a). Anti-idyll: rural horror. In P. Cloke & J. Little (Eds.), *Contested countryside cultures: Otherness, marginalisation and rurality* (pp. 91–104). London: Routledge.

Bell, S. (1997b). Globalisation, neoliberalism and the transformation of the Australian state. *Australian Journal of Political Science, 32*(3), 345–67.

Benn, S. & Dunphy, D. (2007). *Corporate governance and sustainability. Challenges for theory and practice*. New York: Routledge.

Blind, P. K. (2007). *Building trust in government in the twenty-first century: Review of literature and emerging issues*. Paper presented at the 7th Global Forum on Reinventing Government.

Blowfield, M. (2005). Corporate social responsibility - the failing discipline and why it matters for international relations. *International Relations, 19*(2), 173-191.

Bruce, B. (2008, 5th January). At death's door ... but fighting back. *The New Zealand Herald*. Retrieved 4th July 2008, from http://www.nzherald.co.nz/pollution/news/article.cfm?c_id=281&objectid=10485296&pnum=1.

Brueckner, M. (2007). The Western Australian Regional Forest Agreement: economic rationalism and the normalisation of political closure. *Australian Journal of Public Administration, 66*(2), 148–58.

Brueckner, M. (2008). *Balancing people, place and prosperity: Lessons from Western Australia*. Paper presented at the Symposium: Wealth and prosperity in the period of global transformation: experience of Russia and Asia.

Brueckner, M. (2009). *Openness in the face of systemic constraints*. Cologne: Lampert Academic Publishing.

Brueckner, M., Duff, J., McKenna, R. & Horwitz, P. (2006). What are they taking us for? The participatory nature of Western Australia's Regional Forest Agreement process. *Australasian Journal of Environmental Management, 13*(1), 6–16.

Brueckner, M. & Horwitz, P. (2005). The use of science in environmental policy: A case study of the Regional Forest Agreement process in Western Australia. *Sustainability: Science, Practice, & Policy [On-line Journal]. URL: http://ejournal.nbii.org/archives/vol1iss2/0412-017.brueckner.pdf, 1*(2), Accessed 5th July 2005.

Brueckner, M. & Mamun, M. A. (in print). Living downwind from CSR: A community perspective on corporate practice. *Business Ethics: A European Review.*

Buggins, A., Sadler, M. & Morfesse, L. (2006, 4th February). Health fears force police out. *The West Australian*, p. 4.

Bureau of Transport and Regional Economics. (2006). *Skill shortages in Australia's regions.* Canberra: Commonwealth of Australia.

Business WA Today. (2009). Alcoa to fight pollution charge. Retrieved 23rd July, 2009, from http://business.watoday.com.au/business/alcoa-to-fight-pollution-charge-20090723-duhn.html

Calhoun, R., Retallack, C., Christman, A. & Fernando, H. (2008). *Meteorological mechanisms and pathways of pollution exposure: Coherent doppler Lidar deployment in Wagerup. Final report.* Tempe, AZ: Arizona State University.

Cameron B. Auxer et al. vs. ALCOA, INC, (2009).

Carpenter, A. (2006). *Brief ministerial statement: Alcoa Wagerup expansion.* Perth: Office of the Appeals Convenor.

Carroll, A. B. (1991). The pyramid of corporate social responsibility: towards the moral management of organizational stakeholders *Business Horizons July-August*, 39–48.

Carroll, J. & Manne, R. (Eds.). (1992). *Shutdown: The Failure of Economic Rationalism and How to Rescue Australia.* Melbourne: Text Publishing Company.

Carson, R. (1962). *Silent Spring.* New York: Penguin Books.

Chartres, M. & Rowland, P. (2004). *Baseline valuation review.* Perth: Megaw & Hogg Incorporating Property Valuation & Consulting Services.

Chetkow-Yanoov, B. (1997). *Social work approaches to conflict resolution.* New York:: Hawthorn Press.

Churches, S. C. (2000). Courts and parliament dysfunctional in review: forest management as a case study of bureaucratic power. *Australian Journal of Administrative Law, 7,* 141–56.

Coleman, W. & Hagger, A. (2001). *Exasperating calculators.* Sydney: Macleay Press.

Community Alliance for Positive Solutions (CAPS) Inc. (2008, 18th December). List of conditions - Letter to Department of Environment and Conservation.

Community Alliance for Positive Solutions (CAPS) Inc. (2009a). About CAPS. Retrieved 3rd March, 2009, from http://www.caps6218.org.au/about.php

Community Alliance for Positive Solutions (CAPS) Inc. (2009b). Letter to Minister for Planning, Culture and the Arts.

Community Alliance for Positive Solutions (CAPS) Inc. (2009c). *Noise regulation 17 application - Alcoa Wagerup refinery. Letter to the Department of Environment and Conservation*. Yarloop: CAPS.

Community Alliance for Positive Solutions (CAPS) Inc. (2009d). The Yarloop bucket brigade. Retrieved 5th May, 2009, from http://www.caps6218.org.au/yarloop_bucket.php

Community Alliance for Positive Solutions (CAPS) Inc. (2010). *Personal communication*. Yarloop.

Community Alliance for Positive Solutions Inc. (2006). *Appeal to Environmental Protection Authority - Wagerup alumina refinery - Increase in production to 4.7 Mtpa; and Wagerup cogeneration plant assessment #1527*. Yarloop: CAPS.

Conacher, A. & Conacher, J. (2000). *Environmental Planning and Management in Australia*. Melbourne: Oxford University Press.

Cook, M. (2003). *Six month report of Yarloop community health clinic*. Yarloop: South West Population Health Unit - Yarloop Community Clinic.

Corporate Knights Inc. & Innovest Strategic Value Advisors Inc. (2007). Global 100. Most sustainable corporations in the world. Retrieved 1st May, 2007, from http://www.global100.org/2007/index.asp

Costanza, R. (1987). Social traps and environmental policy. *BioScience, 37*(6), 407–12.

Costanza, R., Mageau, M., Norton, B. & Patten, B. C. (1998). What is sustainability? In D. J. Rapport, R. Costanza, P. R. Epstein, C. Gaudet & R. Levins (Eds.), *Ecosystem Health* (pp. 231-239). Carlton (VIC): Blackwell Science, Inc.

Cox, E. (Writer) (1995). A truly civil society [Radio]. In ABC (Producer), *The 1995 Boyer Lectures*: Radio National.

Cox, K. (Ed.). (1997). *Spaces of globalization. Reasserting the power of the local*. New York: The Guilford Press.

Croft, J. (2005). *Results of interviews conducted in Yarloop*. Perth: Community Capacity Buidling Branch - Department of Local Government and Regional Development.

Cullen, M. (2002). *Wagerup alumina refinery: health issues*. New Haven: Yale University.

Curtin University of Technology. (2010). Alcoa Research Centre for Stronger Communities. Retrieved 15th January, 2010, from http://strongercommunities.curtin.edu.au/research_new.htm.

Cuthill, M. (2002). Exploratory research: citizen participation, local government and sustainable development in Australia. *Sustainable Development, 10*, 79–89.

Daly, H. E. & Cobb, J. B. (1989). *For the Common Good: Redirecting the Economy Towards the Community, the Environment and a Sustainable Future*. Boston: Beacon Press.

Dames & Moore Consultancy. (1978). *Wagerup alumina project. Environmental review and management programme - Supplement*. Perth: Alcoa Australia.

Davis, G. (1996). *Consultation, Public Participation and the Integration of Multiple Interests into Policy Making*. Paris: Organisation for Economic Co-operation and Development (OECD).

Department of Environment and Conservation. (2006). *Memorandum of understanding between Department of Environment and Conservation and Community Alliance for Positive Solutions for comparison of ambient air sampling and analysis techniques* Perth: DEC.

Department of Environment and Conservation. (2007). Alcoa Wagerup licence renewal application - 6217/10 - Letter to Community Alliance for Positive Solutions.

Department of Environment and Conservation. (2008). *Conditions of license (License Number 6217/12, File Number L80/83)*. Perth: DEC.

Department of Environment and Conservation. (2009a). Wagerup Tripartite Group. Retrieved 4th April, 2009, from http://portal.environment.wa.gov.au/portal/page?_pageid=93,931490&_dad=portal&_schema=PORTAL

Department of Environment and Conservation. (2009b, 23rd January). Wagerup tripartite group - Letter to CAPS.

Department of Environment and Conservation. (2009c). Wagerup Tripartite Group meeting reports. Retrieved 4th April, 2009, from http://portal.environment.wa.gov.au/portal/page?_pageid=93,2384718&_dad=portal&_schema=PORTAL

Department of State Development. (2009). *State agreements*. Perth: Government of Western Australia.

Diesendorf, M. & Hamilton, C. (Eds.). (1997). *Human ecology, human economy. Ideas for an ecologically sustainable future*. St Leonards: Allen & Unwin.

Donoghue, A. M. & Cullen, M. R. (2007). Air emissions from Wagerup alumina refinery and community symptoms: An environmental case study. *Journal of Occupational and Environmental Medicine, 49*(9), 1027-39.

Dovers, S. (2002a). Sustainability: reviewing Australia's progress, 1992-2002. *International Journal of Environmental Studies, 59*(5), 559-71.

Dovers, S. R. (2002b). Policy, science, and community: interactions in a market-oriented world. *Australasian Journal of Ecotoxicology, 8*, 41-4.

Doyle, T. & Kellow, A. (1995). *Environmental politics and policy making in Australia*. Melbourne: Macmillan.

Drew, R. (1997). *An assessment of liquor burning odour emissions at Wagerup*. Melbourne: SHE Pacific Pty Ltd, Safety.

Dryzek, J. (1990). *Discursive democracy: Politics, policy and political science*. Cambridge: Cambridge University Press.

Dryzek, J. S. (1996). Foundations for environmental political economy: The search for homo ecologicus? *New Political Economy, 1*(1), 27-40.

Earles, W. (1999). *Powerless places and placeless powers: The (re)shaping of human services institutions*. University of Western Australia., Perth.

Eckersley, R. (1998). *Measuring progress: is life getting better?* Collingwood, Victoria: CSIRO Publishing.

Eckersley, R. (2001). Economic progress, social disquiet: the modern paradox. *Australian Journal of Public Administration, 60*(3), 89-97.

Edelman. (2009). *Trust barometer*. Sydney: Edelman.

Eisenhuth, S. & McDonald, W. (2007). *The writer's reader. Understanding journalism and non-fiction.* Cambridge: Cambridge University Press.

Elkington, J. (1994). Towards the sustainable corporation: Win-win-win business strategies for sustainable development. *California Management Review, 36*(2), 90–100.

Elkington, J. & Fennell, S. (1998). Partners for sustainability. *Greener Management International, 24*, 48–60.

Environmental Protection Authority. (1978). *Wagerup alumina refinery proposal by Alcoa of Australia limited. Bulletin 50.* Perth: EPA.

Environmental Protection Authority. (2006a). Media Release - EPA Bulletin 1215 -Wagerup Alumina Refinery - Increase in Production to 4.7 Mtpa; and Wagerup Cogeneration Plant. Retrieved 1st May, 2007, from http://www.epa.wa.gov.au/article.asp?ID=2183&area=News&CID=18&Category=Media+Statements

Environmental Protection Authority. (2006b). *Wagerup alumina refinery - increase in production to 4.7 Mtpa; and Wagerup cogeneration plant. Bulletin 1215.* Perth: EPA.

Environmental Protection Authority. (2007). *State of the Environment 2007 Western Australia.* Perth: EPA.

Ferguson, F. (2006, 22nd November). Yarloop exodus. *Harvey Mail,* p. 1.

Ferguson, I., Lavalette, M. & Whitmore, E. (2005). Introduction. In I. Ferguson, M. Lavalette & E. Whitmore (Eds.), *Globalisation, global justice and social work.* London: Routledge.

Flint, J. (2006a, 10th September). Alcoa expands despite toxins. *The Sunday Times,* p. 16.

Flint, J. (2006b, 8th October). Surviving in the shadow of Alcoa. Town fears expansion. *The Sunday Times,* p. 50.

Flint, J. (2006c, 8th September). Tenants sign health forms. *The Sunday Times,* p. 12.

Flint, J. (2007a, 5th August). Erin Brockovich takes on Alcoa. *The Sunday Times.*

Flint, J. (2007b, 13th May). Trying to get away from Alcoa. *The Sunday Times,* pp. 18–9.

Flint, J. (2008, 27th May). Tests confirm town's fears. *The Sunday Times,* p. 24.

Flyvbjerg, B. (1998). *Rationality and power: Democracy in practice. (Trans. S. Sampson).* Chicago: The University of Chicago Press.

Flyvjberg, B. (2001). *Making social science matter: Why social inquiry fails and how it can succeed again.* Cambridge: Cambridge University Press.

Fook, J. (2002). *Social work: Critical theory and practice.* London: Sage Publications.

Forster, P. (1990). *Internal memo.* Perth: Alcoa World Alumina.

Foucault, M. (1980). *Power/knowledge: Selected interviews and other writings, 1972-1977 (C. Gordon, Trans.).* Brighton: Harvester Press.

Foucault, M. (1997). Technologies of the self. In P. Rabinow (Ed.), *Michel Foucault. Ethics: subjectivity and truth* (pp. 223-252). New York: The New Press.

Fox, C. & Miller, H. (1995). *Postmodern public administration: Toward discourse.* Sage Publications.

Frankental, P. (2001). Corporate social responsibility - a PR invention? *Corporate Communications: An International Journal, 6*(1), 12–23.

Freire, P. (1970). *Pedagogy of the oppressed.* New York: Herder and Herder.

Friedman, M. (1970, 13th September). The social responsibility of business is to increase its profits. *New York Times Magazine,* pp. 122-6.

Funtowicz, S. & Ravetz, R. (1991). A new scientific methodology for global environmental issues. In R. Costanza (Ed.), *Ecological Economics.* New York: Columbia University Press.

Galton-Fenzi, B. (1997). *Liquor burning impact assessment (some health issues considered).* Perth: The Healthy Worker Pty. Ltd.

Gare, A. (2002). Human ecology and public policy: overcoming the hegemony of economics. *Democracy and Nature, 8*(1), 131-41.

Giddens, A. (1994). *Beyond Left and Right.* Cambridge: Polity Press.

Giddings, B., Hopwood, B. & O'Brien, G. (2002). Environment, economy and society: fitting them together into sustainable development. *Sustainable Development, 10,* 187-96.

Goldberger, N. R., Tarule, J., Clinchy, B. & Belenky, M. F. (Eds.). (1996). *Knowledge, difference, and power: Essays inspired by women's ways of knowing.* New York: Basic Books.

Goot, M. (2002). Distrustful, disenchanted and disengaged? Public opinion on politics, politicians and the parties: An historical perspective. In D. Burchell & A. Leigh (Eds.), *The Prince's New Clothes: Why do Australians dislike their politicians* (pp. 9-46). Sydney: UNSW Press.

Government and Community Safety Industry Skills Council. (2009). *Government skills Australia - environmental scan.* Adelaide: Government Skills Australia.

Government of Western Australia. (2003a). *Hope for the future: The Western Australian state sustainability strategy.* Perth: Department of the Premier and Cabinet.

Government of Western Australia. (2003b). *Transcript of evidence, 8th September, 2003. Alcoa Wagerup Refinery at Wagerup.* Perth: Standing Committee on Environment and Public Affairs.

Government of Western Australia & Alcoa of Australia Ltd. (1978). *Alumina refinery (Wagerup) agreement and acts amendment act 1978.* Perth: Government of Western Australia.

Grossman, G. M. & Krueger, A. B. (1995). Economic growth and the environment. *The Quarterly Journal of Economics, 110*(2), 353-77.

Hahn, T. (2002). *In the loop. Video-recording.* Bunbury: Centre for Social Research - Edith Cowan University.

Haines, A. (2006, 9th February). Police leave Yarloop. *Mandurah Mail,* p. 20.

Hamann, R. (2009). An academic's perspective on the role of academics in corporate responsibility. In S. Idowu & W. Leal Filho (Eds.), *Professional perspectives of corporate social responsibility* (pp. 347-362). Heidelberg: Springer.

Hamilton, C. (1996). Generational Justice: The marriage of sustainability and social equity. *Australian Journal of Environmental Management, 3*(3), 163-73.

Hamilton, C. (2002). Dualism and sustainability. *Ecological Economics,*

42(1-2), 89–99.
Hamilton, C. (2007). *Scorcher. The dirty politics of climate change*. Melbourne: Black Inc. Agenda.
Hamilton, C. & Maddison, S. (2007). *Silencing dissent. How the Australian government is controlling public opinion and stifling debate*. Crows Nest, NSW: Allen & Unwin.
Hamilton, C. & Saddler, H. (1997). *The Genuine Progress Indicator. A new index of changes in well-being in Australia* (No. 14). Canberra: The Australia Institute.
Handmer, J. W., Dovers, S. R. & Norton, T. W. (2001a). Managing ecosystems for sustainability: challenges and opportunities. In J. W. Handmer, T. W. Norton & S. R. Dovers (Eds.), *Ecology, Uncertainty and Policy. Managing Ecosystems for Sustainability* (pp. 291–303). Harlow, UK: Pearson Education Ltd.
Handmer, J. W., Norton, T. W. & Dovers, S. R. (Eds.). (2001b). *Ecology, uncertainty and policy. Managing ecosystems for sustainability*. Harlow, UK: Pearson Education Ltd.
Hartman, A. (1990). Many ways of knowing. *Social Work, January*, 3–4.
Healthwise. (2004). *Healthwise cancer incidence & mortality study*. Melbourne: Centre for Occupational and Environmental Health - Monash University.
Hemmati, M. (2002). *Multi-stakeholders processes for governance and sustainability – beyond deadlock and conflict*. London: Earthscan.
Henderson, D. (1995). The revival of economic liberalism: Australia in an international perspective. *The Australian Economic Review, 58*(1st Quarter), 59–85.
Hepburn, H. (2007, 17th April). Yarloop's economy dwindles. *Hervey-Leschenault Reporter*, p. 1.
Hinkley, R. (2000). Developing corporate conscience. In S. Rees & S. Wright (Eds.), *Human rights, corporate responsibility: A dialogue* (pp. 287–95). Annandale: Pluto Press Australia.
Hinwood, A. (2007). *EPA media statement - EPA reveals Western Australia's environment report card*. Perth: EPA.
Hobbs, A. (2008, 15th September). Lobby groups urge Barnett to get moving. *The West Australian*.
Holman, D. (2002). *Summary and recommendations*. Perth: The Wagerup Medical Practitioners' Forum.
Holman, D. (2008). *The Wagerup and surrounds community health survey*. Perth: Telethon Institute for Child Research.
Holman, D., Harper, A., Somers, M., Galton-Fenzi, B. & Phillips, M. (2005). *Wagerup refinery Unit Three Expansion - Letter to the Environmental Protection Authority*. Perth: Independent Members of the Wagerup Medical Practitioners' Forum.
Holman, H. R. & Dutton, D. B. (1978). A case for public participation in science policy formation and practice. *Southern California Law Review, 51*, 1505–34.
Holmes, D. (2008). *The Wagerup and surrounds community health survey*. Perth: Telethon Institute for Child Research.
Holmes, D. & Girardi, G. (1999). *Survey of productivity and environmental*

strategies in Australian manufacturing companies. Melbourne: Australian Industry Group and Monash Centre for Environmental Management.

hooks, b. (1994). *Outlaw culture: Resisting representations.* New York: Routledge.

hooks, b. (2000). *All about love.* London: New Visions.

Hughes, O. (1980). Bauxite mining and jarrah forests in Western Australia. In R. Scott (Ed.), *Interest groups and public policy: Case studies from the Australian states* (pp. 170-93). Melbourne: Macmillan.

Hunold, C. & Young, I. M. (1998). Justice, democracy and hazardous siting. *Political Studies, 46,* 82-95.

Ife, J. & Tesorieroe, F. (2006). *Community development* (3rd ed.). French's Forest: Longman.

Innes, J. E. & Booher, D. E. (2004). Reframing public participation: strategies for the 21st century. *Planning Theory and Practice, 5*(4), 419-36.

Jaensch, D. (1995). *Parliament, politics and people: Australian politics today* (2nd ed.). Melbourne: Longman Australia Pty Ltd.

James, C., Jones, C. & Norton, A. (Eds.). (1993). *A Defence of Economic Rationalism.* Sydney: Allen & Unwin.

Jensen, M. C. (2002). Value maximization, stakeholder theory, and the corporate objective function. *Business Ethics Quarterly, 12*(2), 235-56.

Keith, M. & Pile, S. (1993). *Place and the politics of identity.* London Routledge.

Kellow, A. & Niemeyer, S. (1999). The development of environmental administration in Queensland and Western Australia: Why are they different? *Australian Journal of Political Science, 34*(2), 205-22.

Kelly, B. (1976, 16th September). Refinery site in Hamel area? *South Western Times,* p. 2.

King, S. & Lloyd, P. (Eds.). (1993). *Economic rationalism: Dead end or way forward?* Sydney: Allen & Unwin.

Kok, P., van der Wiele, T., McKenna, R. & Brown, A. (2001). A corporate social responsibility audit within a quality management framework. *Journal of Business Ethics, 31*(4), 285-97.

Korhonen, J. (2002). The dominant economics paradigm and corporate social responsibility. *Corporate Social Responsibility and Environmental Management, 9*(1), 66-79.

Korten, D. (1995). *When corporations rule the world.* San Francisco: Barrett-Koehler Publishers.

Krimsky, S. (1984). Beyond technocracy: new routes for citizen involvement in social risk assessment. In J. C. Petersen (Ed.), *Citizen Participation in Science Policy* (pp. 43-61). Amherst: University of Massachusetts Press.

Kryger, T. (2007). *State economic and social indicators* (No. 14 2007-08). Canberra: Commonwealth of Australia.

Lafferty, W. M. & Meadowcroft, J. (Eds.). (2000). *Implementing Sustainable Development. Strategies and Initiatives in High Consumption Societies.* Oxford: Oxford University Press.

Lagan, A. (2000). *Why ethics matter: Business ethics for business people*. Melbourne: Information Australia.

Laird, F. (1993). Participatory analysis, democracy, and technological decision making. *Science Technology & Human Values, 18*(3), 341–61.

Langley, G. (1976, 23rd September). $1000 mil. refinery plan for Wagerup!, *Western Herald*, p. 1.

Langmore, J. & Quiggin, J. (1994). *Work for all*. Melbourne: Melbourne University Press.

Lantos, G. P. (2001). The boundaries of strategic corporate social responsibility. *Journal of Consumer Marketing, 18*(7), 595–630.

Lawson, R. (2009, 14th September). Gorgon approved, valued at $43bn. *WA Business News*, from www.wsbusinessnews.com.au/story/1/75230/gorgon-approved-valued-at-43bn.

Leonard, P. (1997). *Postmodern welfare: Reconstructing an emancipatory project*. London Sage Publications.

Lines, W. J. (2006). *Patriots : defending Australia's natural heritage*. St. Lucia, Qld: University of Queensland Press.

Lippmann, M. (Ed.). (2009). *Environmental toxicants: human exposures and their health effect*. Hoboken, N.J.: Wiley & Sons.

Llewellyn, P. (2008, 15th May). *Parliamentarians turn their backs on Yarloop residents - Media release*. Perth: The Greens.

Marsden, C. (2000). The new corporate citizenship of big business: part of the solution to sustainability. *Business and Society Review, 105*(1), 9–25.

Mayman, J. (2002, 11th-12th May). The stink of Uncle Al. *The Weekend Australian*, p. 19.

Mazur, J. (2000). Labor's new internationalism. *Foreign Affairs, 79*(1), 79-93.

McChesney, R. W. (1999). Introduction. In N. Chomsky (Ed.), *Profit over people: Neoliberalism and global order.* (pp. 7–18). New York: Seven Stories Press.

McDermott, Q. (2005, 3rd October). Something in the air - ABC Four Corners.

McGowan, M. (2006). *Environmental approval for the Alcoa expansion. Statement by the Minister for the Environment*. Perth: Government of Western Australia.

McKenna, R. & Tsahuridu, E. (2001). Must managers leave ethics at home? Economics and moral anomie in business organisations. *Reason in Practice, 1*(3), 67–76.

Meppem, T. & Gill, R. (1998). Planning for sustainability as a learning concept. *Ecological Economics, 26*, 121–137.

Mercer, A. (2001). *Report on Wagerup health survey*. Perth: Department of Public Health - University of Western Australia.

Mercer, D. & Marden, P. (2006). Ecologically sustainable development in a 'quarry' economy: One step forward, two steps back. *Geographical Research, 44*(2), 183–203.

Mills, C. W. (1959). *The sociological imagination*. New York: Oxford University Press.

Minister for Environment; Youth. (2009, 26th August). *Letter to Community Alliance for Positive Solutions*.

Minister for Planning; Culture and the Arts. (2009, 4th June). Positive solutions for Yarloop and districts - Letter to the Community Alliance for Positive Solutions.

Mitchell, R. K., Agle, B. R. & Wood, D. J. (1997). Toward a Theory of Stakeholder Identification and Salience: Defining the Principle of Who and What Really Counts. *The Academy of Management Review, 22*(4), 853–86.

Mitroff, I. (1998). *Smart thinking for crazy times: the art of solving the right problem.* San Francisco: Berrett-Koehler.

MMSD Project. (2002). *Breaking new ground.* London: Earthscan.

Mol, A. P. J. & Sonnenfeld, D. A. (2000). Ecological modernisation around the world: An introduction. *Environmental Politics, 9*(1), 1–14.

Mol, A. P. J. & Spaargaren, G. (2000). Ecological modernisation theory in debate: A review. *Environmental Politics, 9*(1), 17–49.

Monash University & University of Western Australia. (2002). *Healthwise cancer and mortality study, first report.* Melbourne & Perth: MU & UWA.

Morton, A. (2009, 27th November). Secrecy on aluminium subsidies to remain *The Age*, from http://www.theage.com.au/environment/secrecy-on-aluminium-subsidies-to-remain-20091126-juon.html.

Murphy, J. (2000). Ecological modernisation. *Geoforum, 31*, 1–8.

Murray, P. (1976, 11th September). Big S.W. land deal mystery. *The West Australian,* p. 1.

Musk, A. W. & de Klerk, N. H. (2000). *Health effects from liquor burning unit emissions in an alumina refinery.* Perth: University of Western Australia.

Nelson, N. & Wright, S. (1995). *Power and Participatory Development: Theory and Practice.* London: Intermediate Technology.

Nevile, J. W. (1997). *Economic rationalism. Social philosophy masquerading as economic science. Background Paper No. 7.* Canberra: The Australia Institute.

Nicol, T. (2006). WA's mining boom: Where does it leave the environment? *ECOS, 133*(Oct-Nov), 12–13.

Nobbs, J. (2003). *Yarloop public rental housing. Letter to K.J. Leece, Chief Executive Officer with the Department of Housing and Works.* Bunbury: Homeswest.

Norgaard, R. (Ed.). (1994). *Development betrayed: The end of progress and a coevolutionary visioning of the future.* London: Routledge.

O'Riordan, T. & Stoll-Kleemann, S. (2002). Deliberative democracy and participatory biodiversity. In T. O'Riordan & S. Stoll-Kleemann (Eds.), *Biodiversity, Sustainability and Human Communities. Protecting Beyond the Protected* (pp. 87–112). Cambridge: Cambridge University Press.

Office of Energy. (2003). *Energy Western Australia.* Perth: Government of Western Australia.

Orchard, L. (1998). Managerialism, Economic Rationalism and Public Sector Reform in Australia: Connections, Divergences, Alternatives. *Australian Journal of Public Administration, 57*(1), 19–32.

Orlitzky, M., Schmidt, F. L. & Rynes, S. L. (2003). Corporate social and financial performance: A meta-analysis. *Organization Studies, 24*(3),

403–41.

Osborn, W. (Ed.). (2003). *Mr Wayne Osborn's opening statement to the committee, September 8 2003*. Perth.

Painter, M. (1996). Economic policy, market liberalism and the 'end of Australian politics'. *Australian Journal of Political Science, 31*(3), 287–99.

Pattenden, C. (2004). Sustaining community in a closed mining town. In D. Woods & G. Pass (Eds.), *Alchemies: Community exChanges*. Perth: Black Swan Press.

Pearse, G. (2007). *High and dry: John Howard, climate change and the selling of Australia's future*. Camberwell, Vic.: Penguin.

Pell, D. J. (1996). The Local Management of Planet Earth: Towards a 'Major Shift' of Paradigm. *Sustainable Development, 4*, 138–48.

Pemberton, J. (1988). 'The end' of economic rationalism. *Australian Quarterly, 60*, 188–99.

Pettman, J. (1992). *Living in the margins: Racism, sexism and feminism in Australia*. North Sydney: Allen & Unwin.

Phillips, H. C. J. & Kerr, L. (2005). Western Australia. *Australian Journal of Politics & History, 51*(2), 297–302.

Pile, S. & Keith, M. (1997). *Geographies of resistance*. London Routledge.

Pilger, J. (2006). *Freedom next time*. London: Bantam Press.

Porter, M. E. & Kramer, M. R. (2006). Strategy and society: The link between competitive advantage and corporate social responsibility. HBR Spotlight. *Harvard Business Review, December*, 1–14.

Portes, A. (1998). Social capital: Its origins and applications in modern sociology. *Annual review of sociology, 24*, 1–24.

Pusey, M. (1991). *Economic rationalism in Canberra: a nation-building state changes its mind*. Cambridge: Cambridge University Press.

Pusey, M. (2003). *The experience of middle Australia: the dark side of economic reform*. Cambridge: Cambridge University Press.

Raffensperger, C. & Tickner, J. A. (Eds.). (1999). *Protecting public health & the environment*. Washington, DC: Island Press.

Randal, A. (2008). Is Australia on a sustainability path? Interpreting the clues. *The Australian Journal of Agricultural and Resource Economics, 52*, 77–95.

Rees, S., Rodley, G. & Stilwell, F. (Eds.). (1993). *Beyond the market. Alternatives to economic rationalism*. Leichhardt, NSW: Pluto Press.

Reich, R. (2008). *Supercapitalism. The transformation of business, democracy and everyday life*. Melbourne: Scribe.

Renn, O. (1992). Risk communication: towards a rational discourse with the public. *Journal of Hazardous Materials, 29*, 465–519.

Reputex. (2003). *Social responsibility ratings 2003*. Melbourne: Reputex.

Ritchie, I. (1998, November). *The Changing Face of Extractive Metallurgy*. Paper presented at the Australian Academy of Technological Sciences and Engineering Symposium: Technology - Australia's future: New technology for traditional industries, Tokyo.

Robertson, A. (2005). *Submission by the Department of Health - Wagerup refinery unit three expansion. Environmental review management programme (ERMP)*. Perth: Department of Health.

Ross, D. (2002a). *Enacting my theory and practice of an ethic of love in social work education*. Bunbury: Edith Cowan University.

Ross, D. (2002b). *Yarloop - Alcoa at the crossroads – naming the issues*. Bunbury: Centre for Social Research - Edith Cowan University.

Ross, D. (2003a). *The business case for an ethic of love: The Alcoa/ECU partnership – informing theoretical and ethical constructs*. Bunbury: Centre for Regional Development & Research - Edith Cowan University.

Ross, D. (2003b). *Land management 2001-2003. Initial briefing paper*. Bunbury: Edith Cowan University.

Ross, D. (2003c). *Reviewing the land management process: Some common ground at a point in the process. A report on the collaboration between Alcoa, Wagerup and Yarloop/Hamel property oweners*. Bunbury: Edith Cowan University.

Ross, D. (2007). A dis/integrative, dialogic model for social work education. *Social Work Education, 26*(5), 1–5.

Ross, D. (2009). Emphasizing the 'social' in corporate social responsibility: A social work perspective. In S. Idowu & W. Leal Filho (Eds.), *Professional perspectives of corporate social responsibility* (pp. 301–18). Heidelberg: Springer.

Sandman, P. (2002). Laundry list of 50 outrage reducers. Retrieved 7th November, 2003, from http://www.psandman.com/col/laundry.htm

Saul, J. R. (1997). *The unconscious civilisation*. Ringwood (VIC): Penguin Books Australia Ltd.

Saul, J. R. (2005). *The collapse of globalism and the reinvention of the world*. New York: The Overlook Press.

Schur, B. (1985). *Jarrah forest or bauxite dollars?: a critique of bauxite mine rehabilitation in the jarrah forests of southwestern Australia Perth (WA)*. Perth: Campaign to Save Native Forests (WA).

Singleton, J. & Elliot, P. (2003). *Comments on the review report (June 2003) of Alcoa's revised proposal (January, 2002). Correspondence provided to the Land Management Meetings by senior consultants of URS*. Perth: URS.

Sleeman Consulting & Goodall and Business and Resource Management. (2004). *Energy for minerals development in south west coast region of Western Australia*. Perth: Western Australian Department of Industry and Resources.

Smith, D. E. (1990). *The conceptual practices of power: A feminist sociology of knowledge*. Boston: Northeastern University Press.

Smith, L. T. (1999). *Decolonizing methodologies: Research and indigenous peoples*. London: Zed Books.

Smith, S. (2002, 9th June). Alcoa's bad odour. *Radio National: Background Briefing*.

Southwell, M. (2001, 29th November). Cancer secret. *The West Australian*, pp. 1, 3.

Southwell, M. (2002a, 27th June). Alcoa cancer rate alarm. *The West Australian*, p. 3.

Southwell, M. (2002b, 24th May). Alcoa memo lists problem. *The West Australian*, p. 7.

Stacey, R. D. (1993). *Strategic management and organisational dynamics*. London: Pitman.

Standing Committee on Education and Health. (2007). *Inquiry into the cause and extent of lead pollution in the Esperance area* Perth: Western Australia. Parliament. Legislative Assembly.

Standing Committee on Environment and Public Affairs. (2004). *Report of the Standing Committee on Environment and Public Affairs in relation to the Alcoa refinery at Wagerup inquiry.* Perth, Western Australia: Legislative Council.

Stanley, L. & Wise, S. (1993). *Breaking out again: Feminist ontology and epistemology* (2nd ed.). London: Routledge.

State of the Environment Advisory Council. (1996). *Australia: state of the environment 1996.* Collingwood: CSIRO Publishing.

State of the Environment Council. (2006). *Australia state of the environment 2006.* Collingwood, Victoria: CSIRO Publishing.

Stolley, G. (2009, 22nd December). Alcoa Wagerup trial date set. *The West Australian.* Retrieved 2nd January 2010, from http://au.news.yahoo.com/thewest/a/-/breaking/6615683/alcoa-wagerup-trial-date-set/.

Survey Research Centre. (2001). *Report on Wagerup health survey.* Perth: Department of Public Health - University of Western Australia.

Sztompka, P. (1999). *Trust: A Sociological Theory.* Cambridge: Cambridge University Press.

Taylor, R. (2007, 6th September). Esperance lead report scathing of DEC. *The West Australian.*

Tennyson, R. & Wilde, L. (2000). *The guiding hand: Brokering partnerships for sustainable development.* Geneva: United Nations Office of Public Information.

Theobald, R. (1997). *Reworking Success : New communities at the end of the millennium.* Gabriola Island, B.C.: New Society Publishers.

Thompson, N. (1998). *Promoting equality: Challenging discrimination and oppression in the human services.* London: Palgrave Macmillan

Tickner, J. A. (2003). Precaution, environmental science, and preventive public policy. *New Solutions: A Journal of Environmental and Occupational Health Policy, 13*(3), 275–82.

Towie, N. (2008, 14th December). Alcoa charged: Criminal negligence case to answer. *Sunday Times* p. 16.

Towie, N. (2009, 15th February). Alcoa let off DEC hook. *The Sunday Times*, p. 36.

Turton, H. (2002). *The aluminium smelting industry: Structure, market power, subsidies and greenhouse gas emissions (Discussion Paper Number 44).* Canberra: The Australia Institute.

United Nations - General Assembly. (2001). *Draft resolution introduced on ways to promote peace and progress.* New York: UN.

United Nations Development Programme. (1997). *Reconceptualising Governance. Discussion Paper 2.* New York: UNDP.

United Nations Division for Sustainable Development. (1992). *Agenda 21.* Rio de Janeiro: United Nations.

Utting, K. (2002, 19th March). People we do have a problem – Alcoa. *Harvey Reporter.*

Utting, K. (2006a, 24th October). Hospital close after 111 years. *Harvey-Leschenault Reporter,* p. 1.

Utting, K. (2006b, 14th February). Staff shift a blow for Yarloop. *Harvey-Leschenault Reporter,* p. 1.

Valentine, T. (1996). Economic rationalism vs the entitlement consensus. *Policy*(Spring), 3-10.

van Bavel, R. & Gaskell, G. (2004). Narrative and systemic modes of thinking. *Culture and Psychology, 10*(4), 418-39.

WA Legislative Council. (2007). Alumina refinery (Wagerup) agreement and acts amendment act 1978 - variation agreement - disallowance (pp. 3176g-84a / 3171). Perth: Parliament of Western Australia - Hansard.

WA Parliamentary Debates - Hansard. (2006). Standing committee on environment and public affairs - eleventh report - Alcoa refinery at Wagerup inquiry - Motion (pp. 2038a-42a). Perth: Parliament of Western Australia - Hansard.

Wagerup Medical Practitioners' Forum. (2005). *Submission on ERMP: Wagerup refinery unit three expansion*. Perth: Wagerup Medical Practitioners' Forum.

Walker, B., Carpenter, S., Anderies, J., Abel, N., Cumming, G., Janssen, M., et al. (2002). Resilience management in social-ecological systems: a working hypothesis for a participatory approach. *Conservation Ecology, 6*(1), 14.

Walker, C. (2002a). *On the valuation of properties. Wagerup alumina refinery & Yarloop townsite. III*. Roleystone: Geo & Hydro Environmental Management Pty Ltd.

Walker, K. J. (2001). Uncertainty, epistemic communities and public policy. In J. W. Handmer, T. W. Norton & S. R. Dovers (Eds.), *Ecology, Uncertainty and Policy. Managing Ecosystems for Sustainability* (pp. 262-90). Harlow, UK: Pearson Education Ltd.

Walker, K. J. (2002b). Uncertainty, epistemic communities and public policy. In J. W. Handmer, T. W. Norton & S. R. Dovers (Eds.), *Ecology, Uncertainty and Policy. Managing Ecosystems for Sustainability* (pp. 261-90). Harlow, UK: Pearson Education Ltd.

Walkington, P. (2005). *Environmental Protection Authority briefing note: Site visit on 7 July 20056 for the Alcoa Wagerup refinery unit 3 expansion*. Perth: EPA.

Wallich, H. C. & McGowan, J. J. (1970). Stockholder interest and the corporation's role in social policy. In W. J. Baumol (Ed.), *A new rationale for corporate social policy* (pp. 39-59). New York: Committee for Economic Development.

Washburn, J. (2005). *University Inc. The corporate corruption of Amercian higher education*. Cambridge, MA: Basic Books.

Watson, I., Buchanan, J., Campbell, I. & Briggs. (2003). *Fragmented futures: new challenges in working life*. Annandale, NSW: Federation Press.

Welker Environmental Consultancy. (2003). *Western Australian licence conditions. Independent strategic review. Final report*. Perth: Department of Environmental Protection.

Western Australian Forest Alliance. (2007). Boycottt jarrah. Retrieved 5th May, 2007, from http://www.wafa.org.au/actions/boycottjarrah.html

Whitmore, E. & Wilson, M. (2005). Popular resistance to global corporate rule: the role of social work (with a little help from

Gramsci and Freire). In I. Ferguson, M. Lavalette & E. Whitmore (Eds.), *Globalisation, global justice and social work* (pp. 189–206). London: Routledge.

Wilson, N. (2006, 22nd July). Can-do Alcoa pays a generous social dividend for new alumina plant. *The Australian*.

World Commission on Environment and Development. (1987). *Our Common Future*. Oxford: Oxford University Press.

Wynne, T. (2009). *Letter to the Community Alliance for Positive Solutions Inc*. Perth: Department of Environment and Conservation.

Yarloop and Districts Concerned Residents' Group. (2005). *Submission against the Alcoa Wagerup refinery*.

Young, I. M. (1990). *Justice and the politics of difference*. Princeton: Princeton University Press.

Zadec, S. (2001). *The civil corporation: The new economy of corporate citizenship*. London Earthscan

INDEX

activist achievements 264–5
air quality monitoring 48, 52, 131, 149, 153–4, 187, 191, 201, 266–7
Alcoa (Alcoa World Alumina Australia)
  alumina production, WA 21
  background 20–21
  combative approach 112–13, 137, 244
  community funding by 28, 147, 176
  executive management power 200
  handling of dissent 195–6
  headhunting of government staff 194
  key actions and decisions 249–50
  knowledge of health issues 46
  land management policy document 144
  legal actions against 60, 152–3, 246, 267
  media releases 131, 142, 145, 147, 149, 205, 239–41
  mission statement 143, 242
  science, use in negotiations 118–22, 130–31, 153–7, 199, 235–6, 239
  staff attitudes 118, 131–2, 250
  stated company position 133–6, 140–41
  supporters in Yarloop 83–6, 221, 258
  university funding by 192–4, 227
  values statement 143–4, 242, 257
  see also Wagerup alumina refinery
Alcoa's Centre for Stronger Communities 227
aluminium industry economic loss 39

bauxite mining protests 37, 39, 41
Bayer refining process 44–5
Brockovich, Erin 60
buffer zone 43, 47–50, 55, 102, 116, 134, 147, 173–4, 202, 215, 217, 219–20

Campaign to Save Native Forests (WA) 37
cancers 46, 48
CAPS *see* Community Alliance for Positive Solutions
carcinogens 46
Carpenter, Alan 167, 178, 185, 252
caustic liquor 44, 73, 199
  see also liquor burner
caustic ponds *see* mudlakes
CCN *see* Community Consultative Network
chemical contamination of farms 94
class action writ 60
Community Alliance for Positive Solutions Inc (CAPS) 59–60, 153, 200, 219, 264–8
community consultation process (pre-land management) 32, 34, 53–4, 116, 122–5, 134, 141–2
Community Consultative Network (CCN) 53–4, 146
community funding/ infrastructure 28, 176
community ties 81–2
conflict fatigue 86–7, 267–8
conservationist groups 29, 39, 41
Cookernup 24–5, 37, 41, 48, 61, 217
corporate social responsibility (CSR) 34–6, 246–50, 260, 270
Court, Richard 185
Court, Sir Charles 41, 43, 165, 185
Cowan, Hendy 56
CSIRO 111, 149, 191, 253
CSR *see* corporate social responsibility
Cullen, Mark, Dr 47, 138, 145, 153–7, 241
Curtin University of Technology 193, 227

development, and impacts of 29, 127, 160–63, 176, 179–80, 198, 251–3
see also economic growth
dieback disease 29
dissent, responses to 35–6, 195–6, 253
Donoghue, A.M. 153–7, 241
dust *see under* pollution

economic growth 27–8, 34–5, 41, 159–62, 170, 185–6, 188, 201, 242–3, 263, 269
see also development
economic rationalism 160–62, 251–3
ECU *see* Edith Cowan University
Edith Cowan University (ECU) 51, 126–7, 146, 155
land management study (2002) 51, 202, 207–9, 214–235, 241
Edwards, Judy, Dr 54, 168
emissions 21, 23–5, 31, 42, 45–8, 51–2, 54–5, 60, 66–71, 79, 81, 84, 99, 118, 134, 141, 144–5, 149, 152–6, 176, 187–9, 203, 217
control equipment 46, 145, 154, 157, 217
emotional ties 61
employment creation 27–8, 39, 41, 147, 149–51, 159, 203
Enterprise and Learning Centre 141, 150
Environment and Conservation, Department of 30, 52, 54, 60, 146, 166, 169, 172, 190–91, 194, 200, 213, 266–7
Environmental Protection (Noise) Regulations 1997, WA 48
Environmental Protection Authority (EPA) 24, 28, 41–2, 56, 169
EPA *see* Environmental Protection Authority
Esperance 254–5
ethical issues 196–8, 223, 226–7, 241–2, 248, 256–7, 260

forest clearing 29, 37, 39
Friedman, Milton 247–8

Gallop, Geoff, Dr 168, 185
governance 34–6, 259–63, 270
Government, WA State 23, 27–8, 41–4, 57, 159–98
alleged collusion with Alcoa 174–5, 200, 236, 240, 251
distrust of 23, 29, 161–8, 173–83, 192, 223, 237, 261–3
relationship with industry 29–31, 41–2, 54, 159–98, 260–61, 264
technical expertise 118, 189–92
greenhouse gases 29

Hamel 28, 37, 50, 51, 61, 215
Harvey 28, 37, 98
health issues 21, 24–5, 29, 45, 55, 59, 66–7, 144, 153, 154–5, 188, 249
healthy persons 69–71, 81–2
of Alcoa employees 137, 151, 153, 249
symptoms 21, 24–5, 46–8, 66–71, 77, 80, 109–10, 121, 145
symptoms in animals 69–70, 203
health studies 46–8, 120, 138, 149, 153–6
Health, Department of 23, 47, 52, 139, 153, 213
Holman, Darcy, Prof 234
Housing and Works, Department for 98
Howard, John 172, 253
hydrocarbons 45–6

Intergovernmental Agreement on the Environment, Australian 197

Kwinana alumina refinery 44–5, 146–7

land management strategy 22, 25, 43, 48–51, 56–7, 82–3, 91, 99–108, 116, 126, 134, 140, 144,

155, 173–4, 202, 205–6, 209–12, 214–20, 222–3, 228–34, 249–50
lease agreement conditions 91, 98–9, 110, 155
community meetings 206, 214–17, 228–34, 241, 249–50, 256
liquor burner 24, 31, 44–6, 66, 81, 112, 134–5, 153, 188, 203, 249

McGinty, Jim 98
McGowan, Mark 23, 168
MCS *see* multiple chemical sensitivity
media coverage 21, 31, 57–60, 126, 131, 137, 148, 154–5, 195–6, 208, 213, 216, 227, 239–40, 254–5
mudlakes 32, 41, 199, 202–3
multiple chemical sensitivity (MCS) 47, 52, 68, 149, 154, 189

neoliberal policies 34, 160–62, 170, 251, 263
noise *see under* pollution

odours *see under* pollution
Ombudsman 200
oppression 244
Osborn, Wayne 26–7, 28, 43, 133–6, 157

Parliamentary Inquiry 21, 31, 50–53, 58, 126, 136, 157, 176–7, 218, 233–4, 237
perception of risk 154, 156
Police Act 42
pollution 24, 27, 42, 60, 152, 176, 188–9, 203
air *see* emissions
dust 23, 27, 32, 59, 95, 145–7, 152, 244
noise 25, 27, 32, 48–50, 52, 59, 73–4, 84–5, 116, 134, 216–17
odours 27, 32, 47, 49, 51, 59, 73–4, 85, 134, 144, 149, 153, 155–6, 217, 244
visual 217
water 199, 203

power dynamics and mechanisms 122, 128, 130–31, 136, 148, 158, 172, 192, 200, 207, 221–2, 229, 239, 241, 244–6, 256–7, 261, 263
Precautionary Principle 197
proof, onus of 258–9
property purchases *see* land management strategy
property values 24, 49, 57–8, 86, 102–6
psychosomatic illness 47
public silence 254–5

quality of life issues 71–5

regulation, government 168–79, 183–92, 251–2, 255, 260–62, 265
Residue Drying Areas (RDAs) *see* mudlakes
revenue, government 28, 39, 151, 159, 178–80, 188, 243

scientific expertise, independent 192–5, 235–6, 259, 262
self-regulation, industry 170–72, 184, 187, 200, 260
social divisions 50–51
social impacts and costs 22, 25, 39, 59, 73, 134, 146, 160, 204, 206, 219–20, 243, 259–60
social justice 36, 129, 201, 206, 211, 219, 223, 225, 231, 237,242, 255
solutions to conflict 255–70
Southwell, Michael 195–6
South-West Forests Defence Foundation 37
SPPP *see* Supplementary Property Purchase Program
Standing Committee on Environment and Public Affairs 21, 51, 133
see also Parliamentary Inquiry
State Agreement Act 42, 185–6
State Infrastructure Strategy 150–51
Supplementary Property Purchase Program (SPPP) 56–8, 108, 261

sustainability 27–30, 33–6, 61, 127, 146, 176, 200, 204, 242–3, 261, 270
  Alcoa's approach 201–37
  Yarloop personal narratives 61–125
symptoms *see under* health issues

tall stacks 203, 216–17
toxic plumes 68–9, 71, 111, 259
train traffic 73
trust 262–3

visual pollution 217
volatile organic compounds (VOCs) 44, 46, 187

Wagerup alumina refinery
  employees 62, 128, 135, 137, 151, 163, 258
  environmental plans 42
  expansion proposal 23–8, 32, 50, 54–6, 99, 136, 138, 147–9, 152, 156, 168, 175, 223, 234, 246, 249
  history 21, 37–44
  location 37, 38, 42–3, 55
  production volumes 44, 46, 151
  secrecy 39–40
  social capital score 224
  staff attitudes 118, 127–8, 131–2, 200, 205, 250, 256
  technology 44–5
  weather factors 42–3
  see also Alcoa
Wagerup Medical Practitioners Forum 47, 49, 55, 177, 179, 234
Wagerup Sustainability Fund 141, 150, 211–12
Wagerup Tripartite Group (WTG) 54, 267
Waroona 28, 37, 41, 76, 98, 147
water contamination 199, 203
water resources 37, 39
water usage 29, 39–40, 252
Welker Review 54
Western Australian Community Foundation 150

Yarloop
  activist achievements 264–5
  Alcoa employees in 135
  business closures 22, 58, 94, 96–101, 135
  characteristics 18
  community spirit 19–20, 64
  conflicts and disunity 37, 81–8
  conflict fatigue 86–7, 155, 267–8
  CSR issues 35–6
  demographic changes 88–91, 135, 146
  deterioration 64–6, 212
  family and emotional ties 61, 70, 78–80
  family disruption 59, 75–8, 135, 218
  farming properties 94–5
  future, loss of 91–6
  history 19, 61–5
  hospital closure 85, 98, 223
  lifestyle settlers 63–4
  personal narratives 61–125
  police station closure 85, 98
  population 18, 31, 58, 63, 222
  quality of life, decline of 71–5
  relocation proposal 265–6
  services, loss of 96–8, 218
  social divisions 50–51
  social values, decline of 88–90
  train traffic 73
  see also land management *and* property values
Yarloop and Districts Concerned Residents Committee 30, 59, 149, 235
Yarloop Community Clinic 47
Yarloop Sustainability Plan 237

First published in 2010 by
Fremantle Press
25 Quarry Street, Fremantle, Western Australia 6160
(PO Box 158, North Fremantle, Western Australia 6159)
www.fremantlepress.com.au

Copyright text © Martin Brueckner and Dyann Ross 2010
Copyright foreword © Erin Brockovich 2010
Cover image © Alexander Nicholson/Getty Images

This book is copyright. Apart from any fair dealing for the purpose
of private study, research, criticism or review, as permitted under the
Copyright Act, no part may be reproduced by any process without
written permission. Enquiries should be made to the publisher.

Editing: Deb Fitzpatrick, Janet Blagg
Design: Allyson Crimp
Editorial Assistant: Sarah Kiel

 A catalogue record for this book is available from the National Library of Australia

ISBN 9781921361760 (paperback)

Fremantle Press is supported by the Western Australian State
Government through the Department of Cultural Industries, Tourism
and Sport.

www.ingramcontent.com/pod-product-compliance
Lightning Source LLC
Chambersburg PA
CBHW021144160426
43194CB00007B/677